Discovering Biology

A Book of Problems in Vertebrate Zoology with Special Reference to Man

J. J. Head

Oxford University Press 1972

J. J. Head, M.A., Ph.D., B.Sc., A.R.C.S.
was Team Leader responsible for developing examinations in the Nuffield Biology Project, 1963–1966. He is general editor of the *Oxford Biology Readers*.

Oxford University Press, Ely House, London W.1

Glasgow Bombay
New York Calcutta
Toronto Madras
Melbourne Karachi
Wellington Lahore
Dacca
Cape Town Kuala Lumpur
Salisbury Singapore
Ibadan Hong Kong
Nairobi Tokyo
Dar es Salaam
Lusaka
Addis Ababa

Designed by Jillian Haines
Drawings by Oxford Illustrators Ltd

Photoset by BAS Printers Limited, Wallop, Hampshire and printed in Great Britain at the University Press, Oxford, by Vivian Ridler, Printer to the University

Contents

1 Respiration

1–5	7–10	Lavoisier establishes that burning and respiration are essentially the same process, thus laying the corner-stone of physiology.
6–15	10–21	Structure of the chest. Structure and ultrastructure of the lung. Physiology of breathing and gas exchange.
16–17	21–22	Control of breathing.
18–20	22–25	Transport of oxygen in the blood.
21–26	25–29	Tissue respiration; oxygen debt.

2 Blood and the Circulation

27	30	Haemoglobin an efficient oxygen transporter.
28–31	30–35	Pressure in the human circulation; the movement and distribution of blood; cardiac cycle and its change during exercise.
32–36	35–39	Physiology of respiration in diving mammals.

3 Temperature control

37–40	40–43	Mammalian skin structure in relation to temperature control.
41–43	43–46	The hypothalamus is the human thermostat.
44–46	46–47	Effect of surface area/volume ratio on heat loss in some mammals.
47–49	48–50	Heat exchanger mechanisms in muskrat and rabbit.
50–57	50–55	Temperature control in mammals living in hot deserts. 50–53 The Sahara camel, a large mammal. 54–55 The antelope ground squirrel and kangaroo rat, two small mammals.

4 The eye

58–61	56–60	Structure and function of the mammalian retina; retinal adaptation in nocturnal mammals.
62–65	60–63	Accommodation in man and Elasmobranchs. Long and short sightedness and their correction.
66	64	Function of the extrinsic eye muscles.
67	64–66	Visual adaptation in the tarsier.

5 Nutrition

68–73	67–69	Structure of the mammalian head. Tooth structure. Dental characteristics of man, dog and sheep.
74–81	70–75	Structure and function of the mammalian gut. Digestion and absorption. Importance of diffusion. Ultrastructure of the intestine.
82–88	76–77	Bernard establishes the function of the liver in regulating the sugar concentration of the blood.
89	77	Deamination in the liver.
90	77	Essential components of the mammalian diet.
91–96	77–80	Vitamin deficiency diseases. Magendie's controlled feeding technique. Eijkman's work on beri-beri. Hopkins' refinement of controlled feeding with rats.
97–99	80–82	Essential amino acids.
100–101	82–83	Energy yield from various foods and meals.

6 Endocrine glands

102	84–85	Berthold establishes the concept of hormones.
103–104	85–87	The existence and function of secretin is established by Bayliss and Starling. Later, Ivy improves their technique.
105	87	Mering and Minkowski show that the pancreas produces a hormone in whose absence diabetes mellitus develops.
106	88–89	Banting and Best extract insulin.
107	89	Some of the effects of insulin.
108–112	89–94	Thyroid hormones: effects in man and amphibians. TSH.
113	94–95	Loewi establishes the concept of neurohumours.
114	95	Adrenalin.
115–116	95–96	Comparison of nervous and hormonal coordination. Experimental considerations in investigating hormones.

7 Skeleton, muscles and movement

117–119	97–104	Structure of bone. How muscles are attached to bones.
120–121	104–107	Contraction of striped muscle.
122–125	108–112	Skeleton of human shoulder and forelimb. Antagonistic muscle pairs in arm and leg.
126	112–113	Isometric and isotonic contraction.
127	114	Comparison of human pectoral and pelvic girdles.
128–135	114–122	Structure of the human spine; adaptations in different spinal regions.
136–143	122–130	Adaptations in various pentadactyl limbs.

8 Kidney and reproductive system

144	131	Human male and female urinogenital systems.
145–151	132–138	Structure and function of kidney in man. Formation of urine. ADH.
152	138–140	Testis, sperm, semen, ova, internal fertilization.
153	140–142	The human placenta.
154–155	142–144	The foetal blood supply and its changes after birth.
156	144	Growth of the foetus.

Answers to the questions in this book will be found in the Teacher's Guide.

Foreword

A Foreword can take the form of a perfunctory service rendered to the author of a book, who is then placed under obligation to the writer of the Foreword. My case is very different, because it is I who am under obligation to the author, for the honour of inviting me to write the Foreword, and the interest and pleasure of Discovering Biology in a way which makes me envious of students of today. In my day, biology was in a state of chaos, with thousands of facts, like Pirandello's characters, in search of a general principle, of which there were then sadly few.

Darwinian selectionists, Mendelian geneticists, and Lamarckian conveyancers of acquired characters, fought like pugilists in the dark, without Queensberry rules, until Sir Ronald Fisher showed by his experiments, analysis and synthesis, that experimental proof of the fact and efficacy of natural selection was to be found at the heart of genetics, that genes provided exactly the mechanism needed to explain evolution by natural selection of heritable variations, and that Lamarckians had no leg to stand on. It was a classical case of dissolving an opaque precipitate by adding to the solvent more of the solute: research.

Another break-through, by biophysicists and biochemists has, of course, been the elucidation of the structure of DNA and all that it means for replication, reproduction, information-transmission in the genetic code, synthesis in metabolism by encoded enzymes, and differentiation in embryology.

Meanwhile, molecular biology has opened a window that throws light on a problem even more fundamental. There is no such thing as living matter, but there are living systems composed of matter organized and integrated into functional units of increasing degrees of complexity at different levels: the molecular level of metabolism in the cell, the cellular level, the level of organs, and the levels of whole organisms which have increased in complexity and integration as evolution has advanced.

These integrated functional units have been called integrons, an integron being composed of other integrons on a lower level, and itself contributing to the composition of an integron at a higher level. The important thing about integrons is that the properties of matter in an integron at each level are not predictable from those of integrons at lower levels. In living organisms, new and very remarkable properties emerge in integrons, tending to the regulation and protection of the organism in a manner called teleonomic (*not* teleological). The emergence of new properties and functions as a result of increase in integration is a process to which I have given the name of 'bonus principle'.

The significance of integrons and bonus principle is at the root of biology. Scientists believe that the casual postulate can be legitimately and successfully applied to living organisms, which means that analysis of organisms by physical and chemical methods leads to fruitful results, as shown by plentiful examples in this book. There are other people who deny this and want something else to explain the phenomena of life. They give the name of 'reductionists' to scientists who are satisfied in principle with the application of physics and chemistry to the study of life, and who think that biological methods of research (as in experimental embryology and ethology) will ultimately be analysable in terms of the casual postulate applied to emergent properties under the bonus principle. The objectors call themselves 'organicists', claiming that an organism is more than the sum of its parts. What they are really doing is to introduce by the back door the old sterile notion of vitalism which explains nothing and gives no hope of progress in research.

The successful application of physical and chemical techniques to problems of living organisms (as shown in this book) does not mean that life is 'reduced' to physics and chemistry. On the contrary, the emergence of unpredictable properties at high levels of integration *is* biology, and this 'raises' the field of operations of physics and chemistry. The altruistic behaviour of Man, the most precious observable fact in the world, would be no less precious if it had emerged from neuronic organization at the highest integron level, as a bonus.

There is an interesting parallel effect of molecular biology. Fifty years ago, the study of structure was paramount in the field of morphology. Today, interest has returned to focus on structure, at the level of the cell with its membrane, nucleus and envelope, chromosomes, endoplasmic reticulum, ribosomes, Golgi apparatus, and mitochondria; and at the level of molecules whose stereo-chemical shapes play such an important part in their properties of replication, synthesis, regulation of metabolism by feedback, and in immunology when antibodies 'recognize' antigens. Micro-morphology ha come back at the chemical level.

The mention of feedback leads to the subject of cybernetics, whose application to biology has been very fruitful, for it works not only at the cellular level as in regulatory mechanisms such as the lactose operon, but at the organismal level as in the reciprocal alternating control of pituitary and ovarian secretion, and in the maintenance of homeostasis by nervous control of muscular and glandular activities. This book shows many examples of that.

The progress of science during the past three centuries has made its impact on men in the mass through technology, with its labour-saving inventions, from steam engines to safety matches, electric light and aeroplanes, and it attracted more and more students to its study. But the most recent developments, in the form of pollution of nature, thermonuclear bombs, intercontinental ballistic missiles, and unconscionably expensive space probes and lunar exploration, have brought about a revulsion manifested in the decrease in applications for scientific places in universities by students who turn to the humanities. This change of mind is impulsive but irrational. The pollution of nature is due not to excess of technology but to insufficient technology which must turn its attention to avoidance and correction of techniques that lead to pollution. Nuclear energy will be the only hope of mankind when the natural reserves of energy in coal, oil, and gas are exhausted, and they are being used at accelerating rates scandalously more rapid than those at which they were built up. It is not science that is to blame, and in any case Man cannot now do without it. It is the priorities that are wrong, and here it is so much to be wished that those who wield power, politicians and dictators, whether temporal or spiritual, would turn their attention to Discovering Biology. Perhaps, then, science will become generally attractive again.

Applications of biology have been blessings to mankind. Medicine, surgery, veterinary science, genetic production of higher-yielding cultivated plants and domestic animals, responsible control of pests, are all on the credit side, and biology advancing on all fronts presents a picture of which physicists, with their problems, may well feel envious. But what should cause alarm and despondency is Man's withdrawal from the control of his population, a control that is automatic in other animals where ecological conditions and food-supply regulate production. It is scandalous that Man's lack of this control should be deliberately encouraged, by authorities reverting to scholastic reliance on texts, and ignorant of the fact that the biological function of mating is no less vital for the maintenance of a stable family unit as a social matrix than it is for procreation.

Man prides himself on being the only species to have an ethic and a code of morals, but he is also the only species the individual members of which are not averse to perpetrating mass massacres on fellow individuals of their own species. This does not happen in other animals, however well they may be provided with natural means of aggression in the form of tusks, claws, horns, or venom; for natural selection has favoured those varieties whose behaviour does not lead them to destroy their fellows of the same species. Ironically, it may be because man lacks all these structures and is frail, that he has found survival in collective aggression. He must make progress in Discovering Biology to deal with this disgrace.

Finally, biology is fun. In studying it, one must get one's hands dirty. A. V. Hill has said that a great deal of science is in one's hands. It works, problems get solved, accepted general principles are reliable guides, and the result is a satisfaction, perhaps a 'bonus': the sense of wonder. This is what the reader should get from this book.

Sir Gavin de Beer F.R.S., F.S.A.
Membre correspondant de l'Académie des sciences, Institut de France
Kalinga Laureate of UNESCO for popularisation of science

I Respiration

It is difficult for us, living in an age when scientific knowledge is widespread and the amount of scientific research conducted enormous, to understand what it was like to live in an age without science. It is easy to find the notions of our ancestors quaint and amusing, unless we try to imagine the limitations of having no scientific method to help in the understanding of nature. Thus until the beginning of the seventeenth century, the teaching of the Greek anatomist Galen (A.D. 131-210) was accepted as correct, even in an age which produced Michelangelo, Leonardo da Vinci, and Shakespeare. Galen was not aware of the circulation of the blood. He believed the liver to be the centre of the venous system and that the left ventricle received a 'pneuma' (a life-spirit) from the lungs. The right ventricle expelled 'soot', that is excreta, to the lungs, where it was got rid of by exhaling. 'Pneuma' was conveyed to the body from the left ventricle along the arteries and gave the body life.

Until the nineteenth century it was a matter of debate as to whether life was something quite separate from non-living processes. Today no scientist would believe in a supernatural 'life-force' that made non-living things living. All the biological phenomena that have been investigated, though they may be complex, can be explained in terms of mathematics, physics, and chemistry.

The first person to demonstrate clearly that living and non-living processes are the same was the French scientist Lavoisier (1743–94, guillotined at the age of 51 in the name of the Revolution). He concluded that mammalian respiration was essentially the same process as the combustion, or burning, of carbon. His apparatus was crude (though he had a good balance and weighed accurately to three decimal places); but he insisted on quantitative data, and his arguments based on them were essentially the same as scientists would make today.

Like other great scientists he seems to have thought of the answer to his problem first and then set out to test whether he was right. He compared respiration in a guinea-pig with the burning of carbon in two ways:

a the amount of carbon dioxide produced per unit of oxygen consumed in both processes

b the amount of heat given out in the formation of a unit quantity of carbon dioxide in both processes.

Before describing the experiments and asking you questions about them, one last note of the historical background. Until this time the nature of fire was unknown. It was thought by many to be one of the four 'elements' (earth, air, fire, water) of which everything in the world was supposed to be composed. About 1700 a philosopher named Stahl put forward the idea that everything which could burn (e.g. wood, coal, paper) was composed of its own ash (the nature of chemical elements was unknown) together with a substance called *phlogiston*. The word phlogiston comes from a Greek word meaning 'burnt-up' or 'inflammable'. In burning, the phlogiston was supposed to be set free, leaving behind only the ash.

Lavoisier showed that phlogiston was nothing but a figment of the imagination and that burning is a chemical reaction between oxygen and the substance being burned.

Lavoisier had no pure carbon. Instead he used coke.

1 He took a bell jar and stood it in a dish of mercury. Into the jar he introduced as pure oxygen as he had available. It contained 16/19ths (84%) of oxygen and about 1/57th (1·8%) part of carbon dioxide. He does not tell us how he prepared his oxygen, so we cannot know what gases constituted the remaining 14%. On the mercury inside the jar he floated a dish containing a piece of coke. He noted the level of the mercury in the jar.

2 He ignited the coke, which burned for twenty minutes. After allowing the jar to cool to room temperature he noted the new level of the mercury, which had risen slightly.

3 He then introduced caustic potash (KOH) into the jar, which absorbed the carbon dioxide formed by the combustion of the coke. He marked the new level of the mercury, which now almost completely filled the jar.

4 After removing the bell jar he was able, by filling it to the levels he had marked, to calculate the volume of oxygen consumed and the volume of carbon dioxide produced. Since the gas in the jar had been subjected to different pressures when the mercury rose up the jar, he converted all the volumes to what they would have been at atmospheric pressure.

5 His results were that 30·6 g of coke when burned completely, consumed 80·76 l of oxygen and produced 60·42 l of carbon dioxide, i.e. one volume of oxygen produced $\frac{60{\cdot}42}{80{\cdot}76}=0{\cdot}75$ volume of carbon dioxide. 17·4% of his coke remained as ash.

A What light do the results in 5 throw on the phlogiston theory mentioned above?

B The following equation represents the combustion of pure carbon and oxygen:

$C+O_2 \rightarrow CO_2+\text{energy}$

i What volume of carbon dioxide would be produced from one volume of oxygen in this reaction?

ii How did Lavoisier's result differ from this?

iii From the description given of Lavoisier's experiment, suggest why his result differed from that given in i.

iv Assuming that all the oxygen was used up in the combustion with coke, suggest an explanation for the observation made in stage 2 of the experiment.

2

1 Lavoisier set up his apparatus again and passed a live guinea pig through the mercury into the space containing oxygen (the same slightly impure gas as in the last experiment).

2 The animal was allowed to breathe this air for seventy-five minutes and was then removed through the mercury.

3 He marked the new position of the mercury, which had risen a little.

4 He introduced caustic potash (KOH), which caused a considerable rise, and marked the new level of mercury. As before he measured the volumes of gas by filling the bell jar with water to the marks that had been made and adjusted these volumes to what they would have been had the gas in the bell jar been at atmospheric pressure when each mark was made.

5 He found that 0·93 l of oxygen were consumed and 0·76 l of carbon dioxide were produced.

A How many volumes of carbon dioxide were produced for each unit volume of oxygen consumed in respiration?

B i How does this compare with the volume of carbon dioxide produced from one volume of oxygen in the combustion of coke?

ii What arguments can you think of for saying this result does or does not indicate that respiration and combustion are essentially similar processes?

In answering, bear in mind the chemical and quantitative reaction which Lavoisier had demonstrated in respiration and combustion, the errors of the experiment, the phlogiston theory, and the notion that body heat was due to some 'vital force' outside the scope of mathematics, physics, and chemistry.

C These experiments were conducted in 1780. Jumping forward a hundred years or more we can write the following chemical equation:

$C_6H_{12}O_6+6O_2=6CO_2+6H_2O+\text{energy}.$

The ratio of carbon dioxide produced to oxygen used per unit time $\left(\frac{CO_2}{O_2}\right)$ is called the respiratory quotient **or** R.Q.

Here it is $\frac{6}{6}=1$. For the oxidation of fat, the value is:

$C_{57}H_{104}O_6+80O_2=57CO_2+52H_2O$

$R.Q.=\frac{57}{80}=0{\cdot}71$

When protein is converted into respirable substances and respired, the R.Q. obtained is about 0·8. For a mammal on a mixed carbohydrate, fat and protein diet, the R.Q. is around 0·85.

i What value did Lavoisier find for the R.Q. of the guinea pig?

ii Suggest why this measurement was more accurate by modern standards than his similar measurements on coke.

iii Explain the slight decrease in the volume of air in the bell jar noted in stage 3 of the experiment.

3

Lavoisier wanted to measure the amount of heat given out by a burning substance, and devised the apparatus shown in FIGURE 1.1. It was packed with finely crushed ice as shown. On top was fixed a hollow lid, also packed with crushed ice. The hot object was placed in the inner chamber. It melted the crushed ice in the middle chamber. The water produced was collected in the jar beneath.

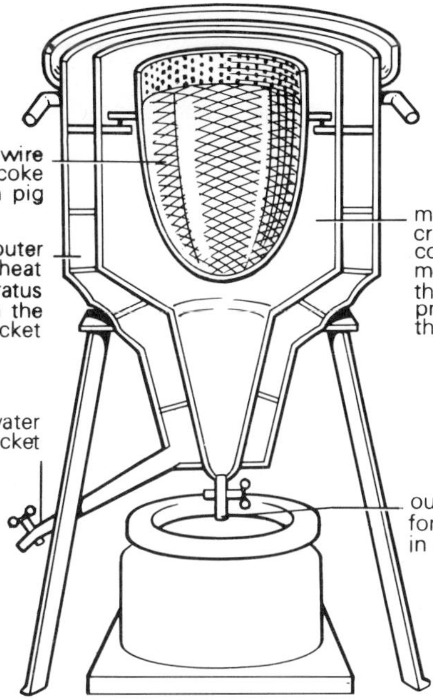

FIGURE 1·1

The volume of melted ice provided a measure of the heat given out by the hot object. The crushed ice in the outer chamber was to prevent heat from the room from melting the ice in the middle chamber; water from it passed out of the side tap.

1 He placed a known weight of glowing coke in the inner chamber and kept it burning by means of bellows. It took 32 minutes to burn away. From the amount of ice melted he calculated that the burning of 30·6g of carbon (again he took coke to be pure carbon) melted 2·998kg of ice. From other experiments he knew that when 30·6g of carbon burned it formed 102·3g of carbon dioxide. From this he calculated that the production of 30·6g of carbon dioxide by the combustion of coke melted 899g of ice.

2 Next he placed a live guinea pig in the inner chamber for ten hours and collected the water from the ice which melted. He ventilated the chamber once or twice and noted that the animal emerged quite happy and having the same body temperature (measured in the rectum) as when it was put in. 402·2g of ice were melted by the animal.

3 Lavoisier already knew the mass of a standard volume of carbon dioxide, and from the measured volumes of gas changes in the experiment in which the guinea pig breathed in a bell jar over mercury (question 2), he calculated that in seventy-five minutes the animal had given out 1·38g of carbon dioxide. Therefore in ten hours it would have produced 11·05g of carbon dioxide.

4 We know from 1 above that the production of 30·6g of carbon dioxide by combustion of coke melts 899g of ice. From this it is possible to calculate how much ice would be melted in the formation of 11·05g of carbon dioxide. It would be 326g.

Lavoisier found in 2 above that the animal in fact melted 402·2g of ice.

A From his experiments on the changes of composition of air after coke had been burned in it and after a guinea pig had respired in it, Lavoisier concluded that combustion and respiration are essentially similar processes.

i How is this conclusion substantiated by the results of the experiments on measuring the amount of heat given off in the two processes?

ii What comment does the result throw on the idea that animal heat is derived from a 'vital force' and is not explicable in terms of mathematics, physics and chemistry? In answering, bear in mind the nature of the experiment and its probable sources of error.

B Lavoisier's own words were 'Thus the heat which is liberated in the conversion of oxygen to carbon dioxide by respiration is the principal source of animal body heat; if other causes are involved, they are of lesser significance'.

Today if we were talking about respiration we would not speak in terms of the heat liberated in the conversion of oxygen to carbon dioxide.

i What chemical reaction would we say had liberated the heat?

ii Why did Lavoisier talk in terms of the heat liberated in the conversion of oxygen to carbon dioxide?

4

Let us examine some of the sources of error in the experiment described in question 3.

A **What error would arise from keeping the coke burning and from ventilating the animal?**

B **Lavoisier observed that burning 30·6 g of coke caused the melting of 2·998 kg of ice. Assuming it was all converted into carbon dioxide, this means that the formation of 30·6 g of carbon dioxide would melt 899 g of ice.**

Why might this assumption not be correct?

C **The argument is that the formation of 30·6 g of carbon dioxide from burning coke melts 899 g of ice. Therefore the formation of 11·05 g of carbon dioxide in animal respiration would require the liberation of sufficient heat to melt 326 g of ice. This assumes that all the energy liberated in respiration appears as heat.**

In the light of modern knowledge, why is this assumption invalid?

5

Lavoisier concluded by suggesting that respiration is exactly similar to the combustion of carbon and that it takes place in the lungs, i.e. that the lungs generate heat which is transferred to the blood passing through the lungs and so is used to keep the animal warm. This combustion occurs 'without any perceptible liberation of light because the fire, as fast as it is freed, is absorbed by the dampness of the lungs'.

How does this speculation match up to present day knowledge of respiration?

6

FIGURE 1.2 is a photograph of part of a human skeleton, showing the chest.

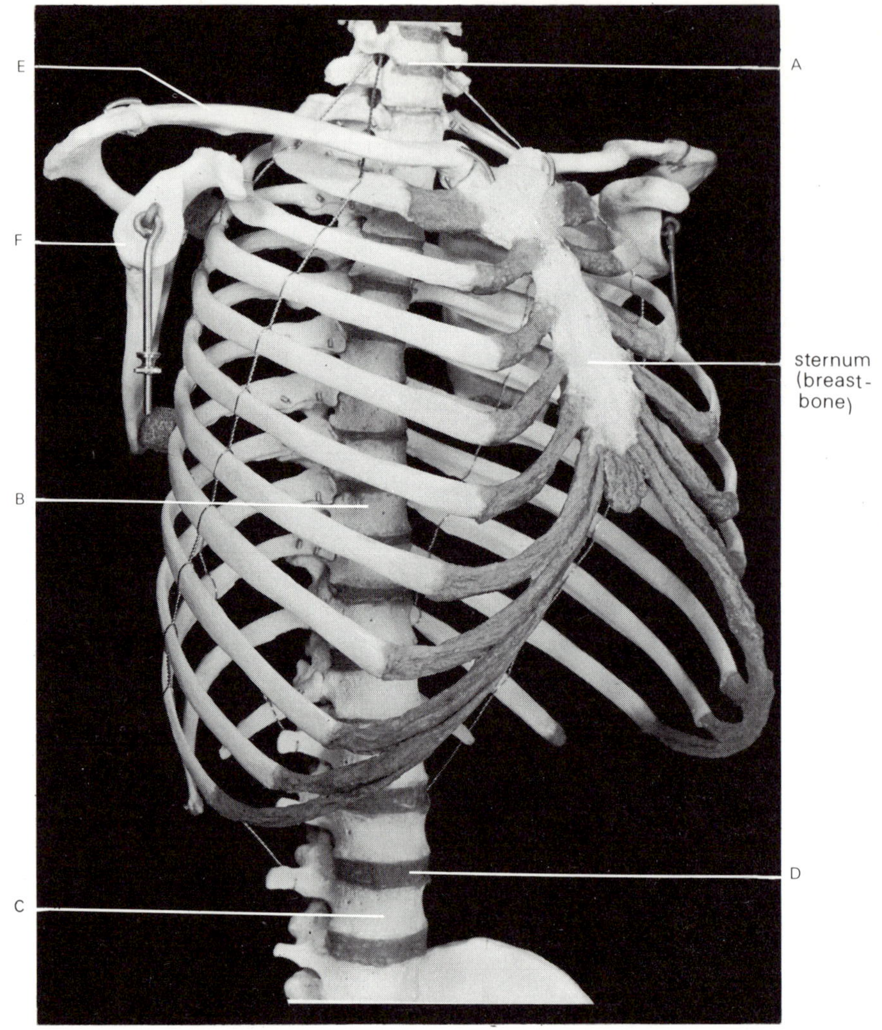

FIGURE 1·2

A **What are the parts labelled** A, B, C, D, E **and** F **?**

B **Explain the sort of body movement that** D **and other similar parts allows.**

C **The ribs can be divided into three types according to the way in which they are attached to the rest of the thoracic skeleton.**

From FIGURE 1.2, **describe the three types, and say which ribs belong to each type.**

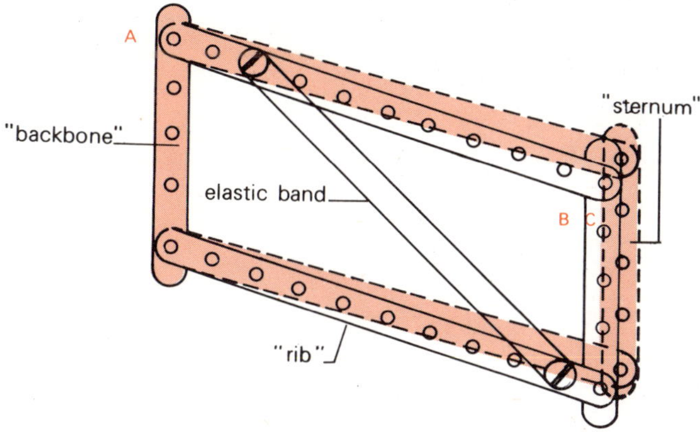

FIGURE 1·3

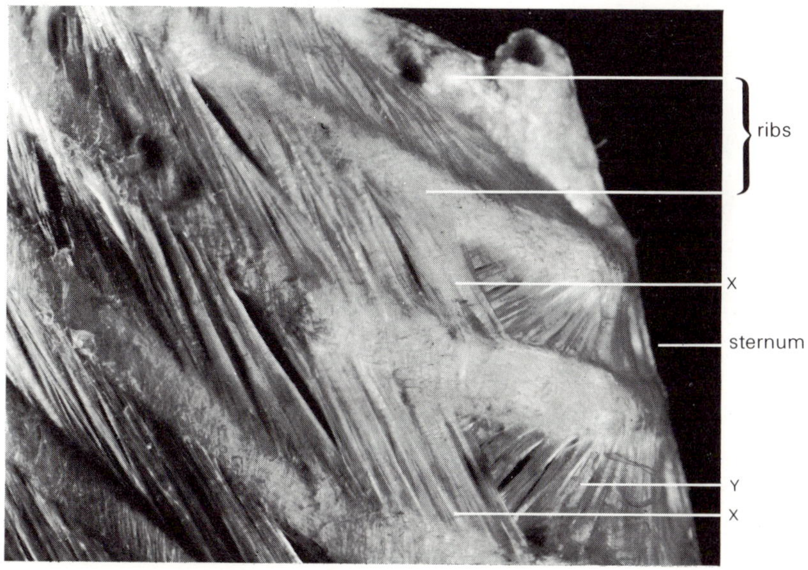

FIGURE 1·4

D **The diaphragm forms the floor of the chest and is attached to the rib cage. Look at** FIGURE 1.2 **and say to which parts of the rib cage you would expect the outer margin of the diaphragm to be attached.**

FIGURE 1.3 is a simple Meccano model to illustrate part of the movement of the chest in inspiration.

If you have not already seen it, make one.

Hold the 'backbone' still and arrange the rest of the model in the position shown. When you release the 'sternum', the 'ribs' rise up to the new position shown by the dotted lines.

A **Explain how the change in the shape of the model, as shown in** FIGURE 1.3, **shows that a similar upward movement of the ribs would increase the volume of the chest.**

B FIGURE 1.4 **is a photograph of part of a human chest with the skin removed. You can see the ribs and sternum but the backbone is not included in the picture.**

- **i** **What are the structures labelled** X **?**
- **ii** **How do these structures bring about a similar movement in the chest to that which you have seen in the Meccano model?**
- **iii** **What, then, will be the effect on the chest of similar activity by the structures labelled** Y **?**
- **iv** **When the model has risen into the position shown by the dotted lines, it has to be pulled down to get it back to its former position. Two main forces operate to make the chest return to its former position during expiration, thus stretching the structures** X. **What are these forces?**
- **v** **Occasionally the structures** Y **are very active. What effect will this have on the chest? At what times does this happen?**

C **The rising of the chest is responsible for the intake of about 40% of each breath. The other 60% is taken in because of the downward movement of the diaphragm.**

Explain:

- **i** **how the muscle fibres of the diaphragm are arranged in relation to the central tendon of the diaphragm and the chest wall;**
- **ii** **how this arrangement causes the diaphragm to be lowered at inspiration;**
- **iii** **how the wall of the abdomen (belly) and the abdominal viscera allow the diaphragm to be lowered, thus expanding the volume of the chest;**
- **iv** **what two principal forces will cause the diaphragm to regain its former position on expiration.**

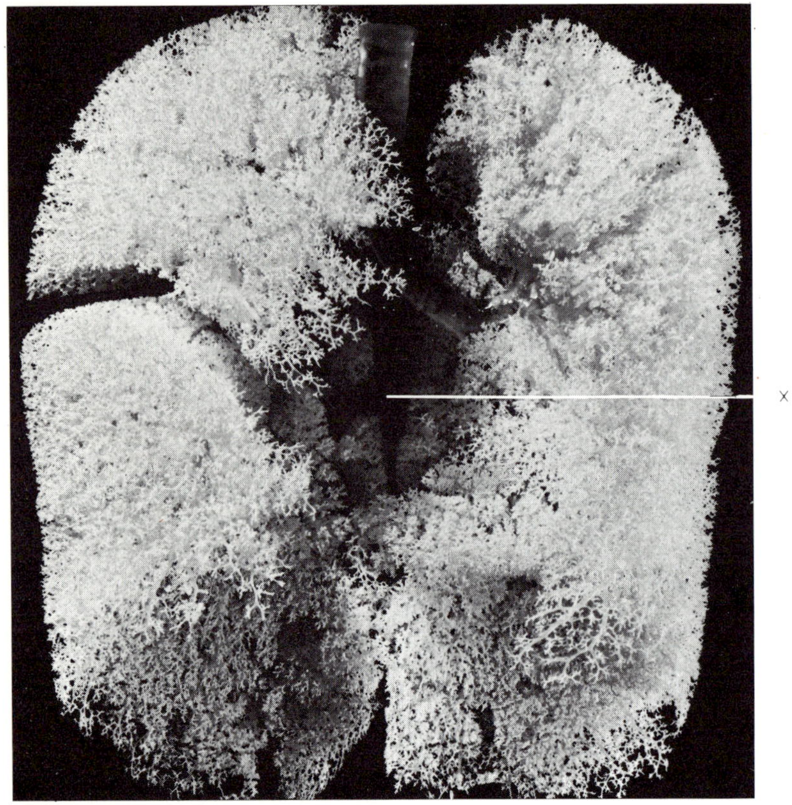

FIGURE 1·5

8

FIGURE 1.5 is a corrosion preparation of the human lung. In making it, plastic material was run into the lung of a corpse so that it penetrated all the spaces of the lung. The plastic set; the lung tissues were dissolved away leaving this cast of the lung.

FIGURE 1.6 is a similar preparation where the end branches (the lobules) have been trimmed off, revealing more detail of the bronchial tree.

A **In life, what is situated in the space marked** x **in** FIGURE 1.5 **?**

B **What is the name of the thin membrane which covers the lung and forms its outer margin?**

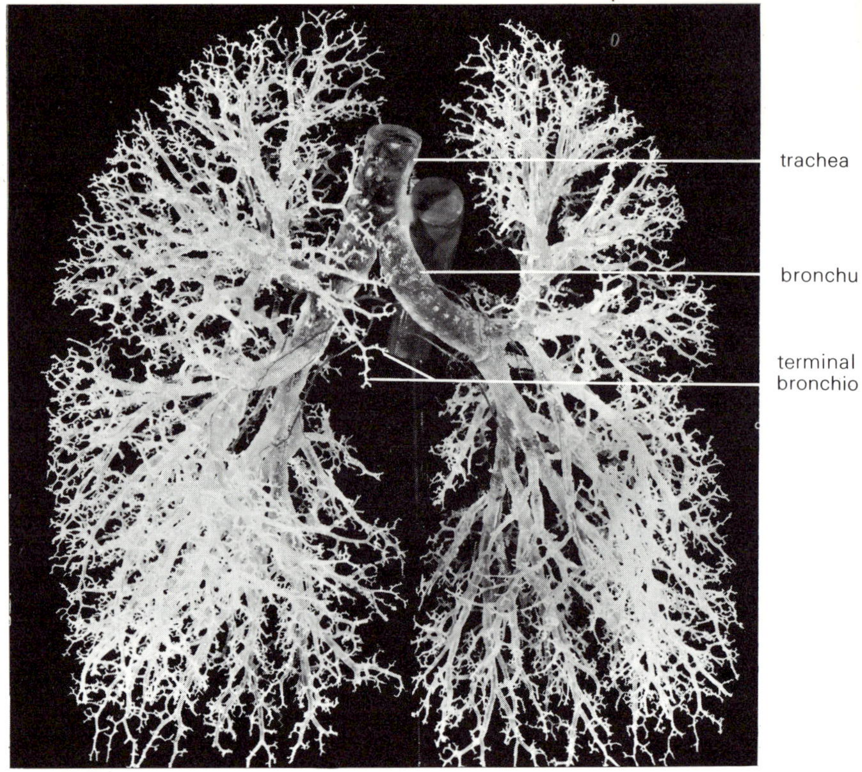

FIGURE 1·6

C **What does** FIGURE 1.5 **tell us about the end branches of the bronchial tree, apart from their shape?**

D **Comparing** FIGURE 1.5 **with** FIGURE 1.6, **what might one say about the relative volume of the end branches (lobules) of the lung compared with the volume of the bronchial tree?**

9

FIGURE 1.7 is a diagram of one of the end branches of the lung (a lobule). The terminal bronchiole shown would be about 1 mm in diameter.

Find this structure in FIGURE 1.6.

You can see that it has arisen from repeated branching of the

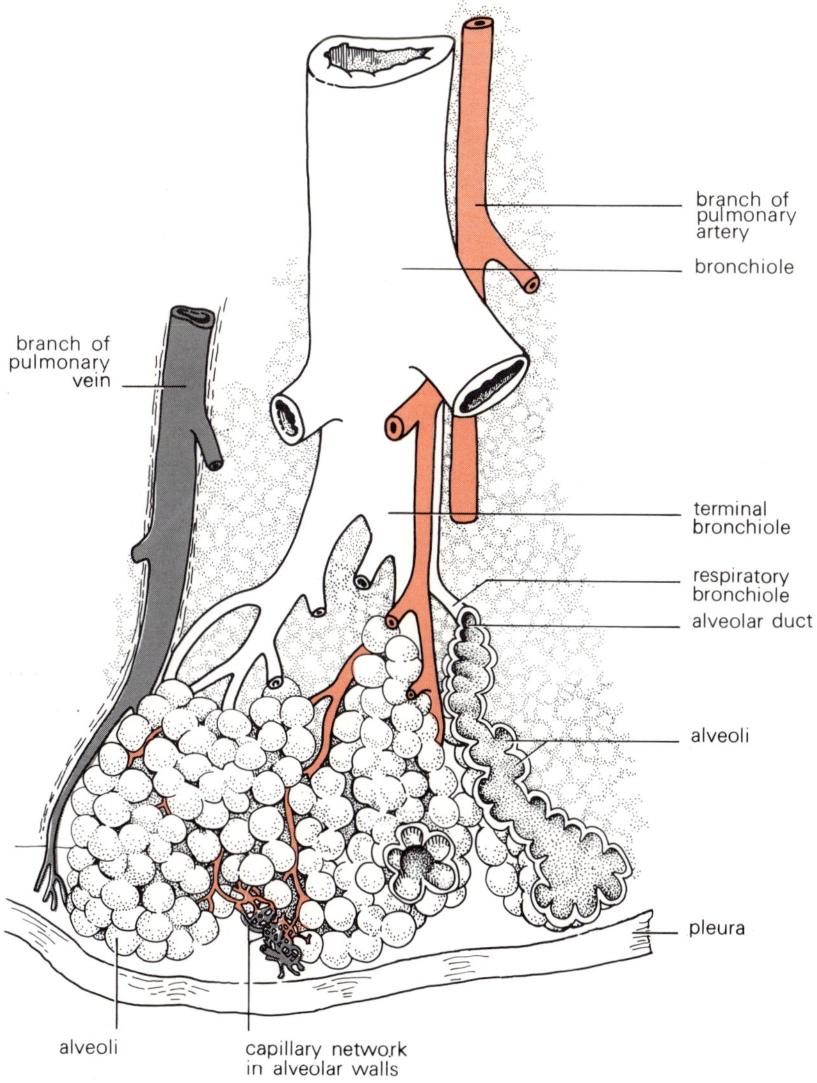

FIGURE 1·7

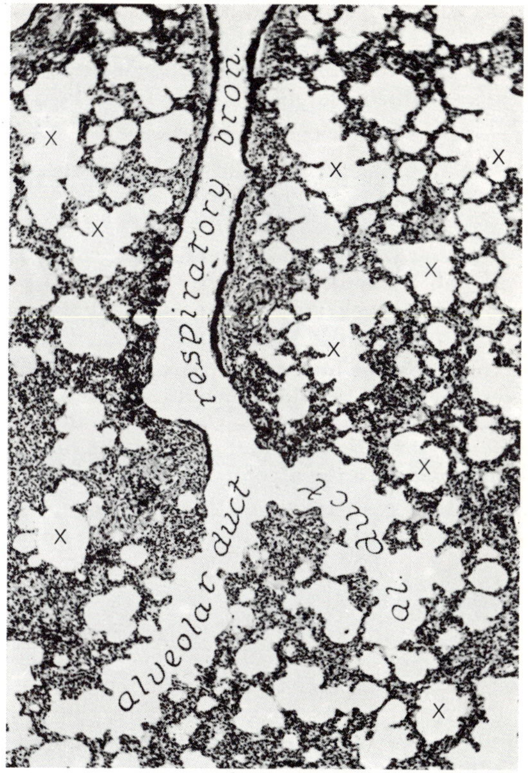

FIGURE 1·8

larger bronchioles which in their turn have arisen from the trachea. The walls of the trachea and the main bronchi contain cartilage (gristle), which supports them and prevents them from collapsing when the pressure in the lung goes down on inspiration.

The walls of the terminal bronchioles, such as those shown in FIGURE 1.7, do not contain cartilage.

The terminal bronchiole divides several times. Each branch is a *respiratory bronchiole* ending in alveolar ducts, the walls of which are, as it were, scooped out into the many cup-like alveoli.

Round the outside of the alveoli lies the capillary network connecting the pulmonary artery to the pulmonary vein.

To summarize: the lung is divided into units called lobules. Each lobule is served by a terminal bronchiole, which divides to form several respiratory bronchioles. Each of these leads into an alveolar duct, the walls of which, the alveoli, look from the outside like a mass of bubbles.

FIGURE 1.8 is a section of a lobule of the lung of a young child, showing a respiratory bronchiole leading to alveolar ducts.

A **Make sure that you can understand how** FIGURE 1.8 **(a two-dimensional section) is related to** FIGURE 1.7 **(a three-dimensional stereogram). What, then, are the structures labelled** X **in** FIGURE 1.8 **?**

B **Gas exchange in the lung takes place in the alveoli, between the alveolar air and the blood passing through the capillaries which lie outside the alveoli. Oxygen is taken up into the blood from the alveolar air; carbon dioxide is released from the blood into the alveoli.**

With regard to gas exchange all the part of the lung shown in FIGURE 1.6, **together with the mouth and pharynx, is referred to as** dead space. **Why do you think it has this name?**

10

FIGURE 1.9 shows the percentage volume of oxygen, carbon dioxide and nitrogen in inspired air (i.e. atmospheric air), expired air (i.e. air as it emerges from the nose) and alveolar air (i.e. the air filling the alveoli).

A **Compare the oxygen content of inspired air with that of expired air. What percentage of the oxygen content of inspired air is taken up into the blood stream?**

B **The amount of oxygen in expired air is 16·3—14·2% = 2·1% more than that in alveolar air; the amount of carbon dioxide in expired air is 5·5—4·0% = 1·5% less than in alveolar air.**

What has happened to the alveolar air on its way from the alveoli that has produced this change in composition?

C **In a single quiet breath, about 500 cm³ of air are breathed in and out of the adult human lung. Assume for simplicity that the alveolar air contains 6% carbon dioxide and that inspired air contains none. By comparing these values with that of 4% for expired air, calculate the volume of the dead space in the adult lung.**

	INSPIRED (ATMOSPHERIC) AIR	EXPIRED AIR	ALVEOLAR AIR
Oxygen	20.96	16.3	14.2
Carbon dioxide	0.04	4.0	5.5
Nitrogen	79.00	79.7	80.3
Totals	100.00	100.0	100.0

FIGURE 1·9

D **What, therefore, is the volume of air entering the alveoli during a quiet breath?**

E **At the end of a quiet breath (when the ribs have descended and the diaphragm has risen) the lungs still contain about 3l of air.**

i **Under these conditions, what is the volume of the dead space?**

ii **How many times greater, therefore, is the volume of the alveolar ducts and the alveoli than the volume of the dead space?**

iii **How does your answer to** ii **fit with the appearance of the corrosion preparations shown in** FIGURES 1.5 **and** 1.6 **?**

F **Thus in a quiet breath about 350 cm³ of air is drawn into the alveolar ducts and the alveoli, whose volume is about 3l.**

How, then, would you expect the composition of alveolar air to vary during a quiet breath?

G **i** **How does the nitrogen composition of expired air compare with that of inspired air** (FIGURE 1·9)**?**

ii **Nitrogen is an inert gas which is not used in the body. While it may to some extent dissolve in the blood there is no diffusion gradient for it into or out of the blood since it is neither used by nor produced by the body tissues.**

In order to explain the higher nitrogen content of alveolar air we must remember that the respiratory quotient of a person on a mixed diet of fat, carbohydrate and protein is about 0·85, i.e.

$$\frac{\text{volume of carbon dioxide produced by the person in unit time}}{\text{volume of oxygen consumed by the person in unit time}} = 0{\cdot}85$$

How does this information explain the relative rise in nitrogen content of alveolar air over that of inspired air?

11

FIGURE 1.10 is another corrosion preparation of the lung. This time the pulmonary artery and pulmonary vein have been injected with plastic. The branches of the pulmonary artery show up light in the picture, those of the vein dark. The trachea, bronchi and bronchioles were dissolved away and do not appear in the preparation.

A i **What may be said of the branching of the pulmonary artery compared with the branching of the pulmonary vein?**

ii **By comparing** FIGURE 1.10 **with** FIGURE 1.6, **what may be said of the branching of the pulmonary artery and the pulmonary vein compared with the branching of the bronchial tree?**

iii **Comparing** FIGURE 1.10 **with** FIGURE 1.5, **what is conspicuously missing from the corrosion preparation of the blood supply to the lungs?**

iv **How can this be confirmed by inspection of** FIGURE 1.10**?**

12

FIGURE 1.11 is another preparation of blood vessels from the lung. The blood vessels were injected with an opaque material, and this is a high power photomicrograph of a small part of one lobule.

Refer back to FIGURE 1·7 **and say what you think this photograph shows. What are the portions labelled** X **?**

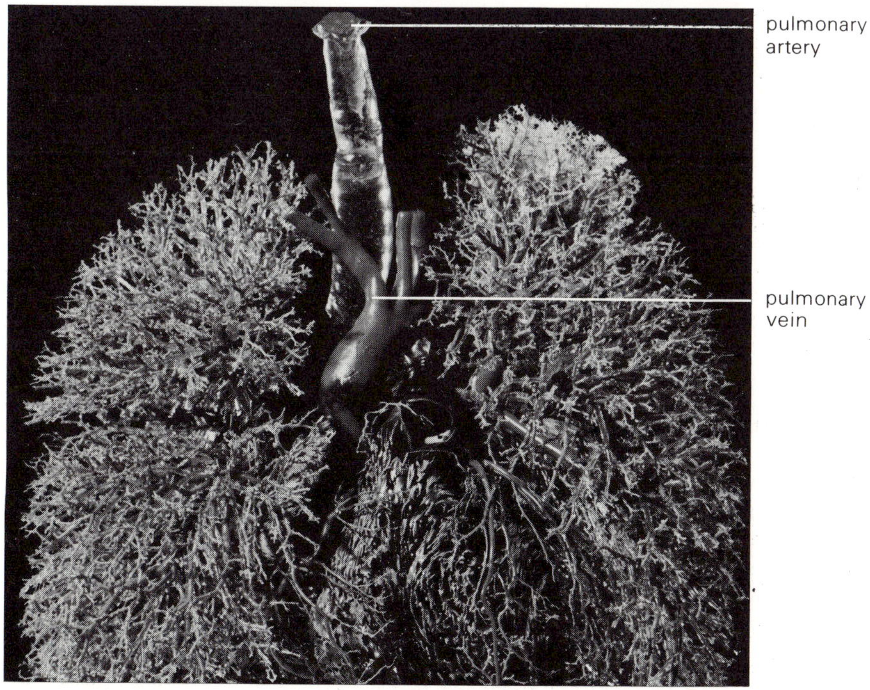

FIGURE 1·10

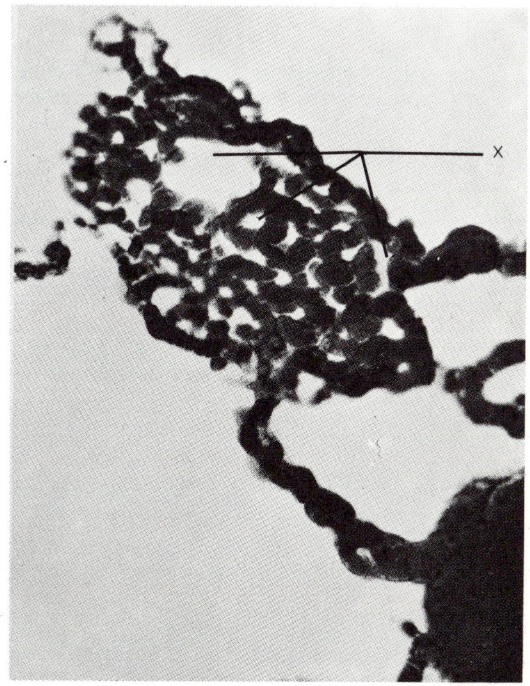

FIGURE 1·11

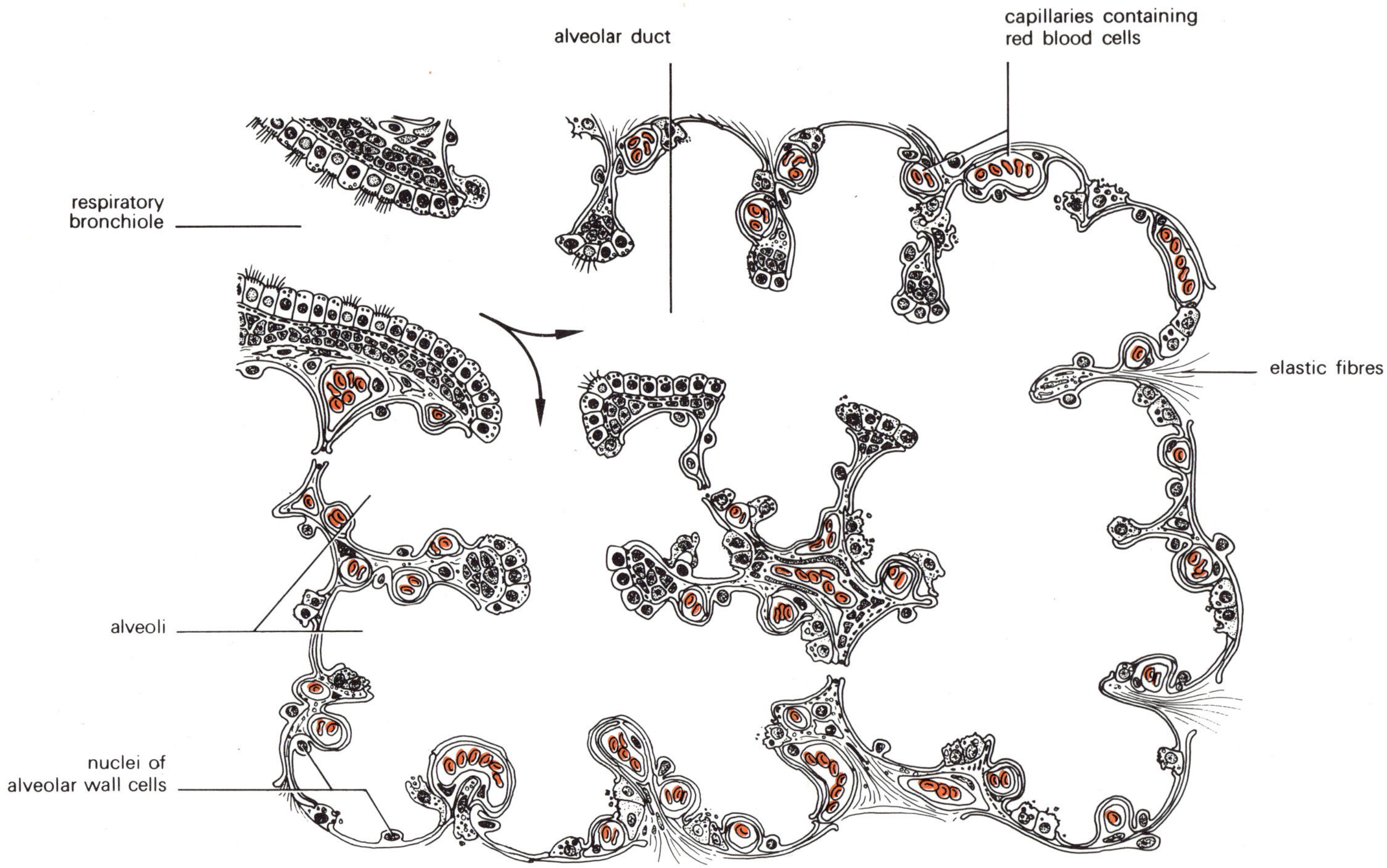

FIGURE 1·12

13

We are now going to look at the structure of the alveoli. FIGURE 1.12 is a drawing of a section through a respiratory bronchiole leading into two alveolar ducts, which in turn lead into the alveoli. Some of the capillaries associated with this lobule are shown cut in transverse section; the red blood cells in them are drawn in red.

Make sure that you understand how FIGURE 1.12 is related to FIGURE 1.8 (a photograph of a section of lung) and to FIGURE 1.7 (a stereogram of part of one lobule).

A In the rather crude drawing of FIGURE 1.7, it appears that the capillary network lies on top of the alveolar wall. Look carefully at FIGURE 1.12.

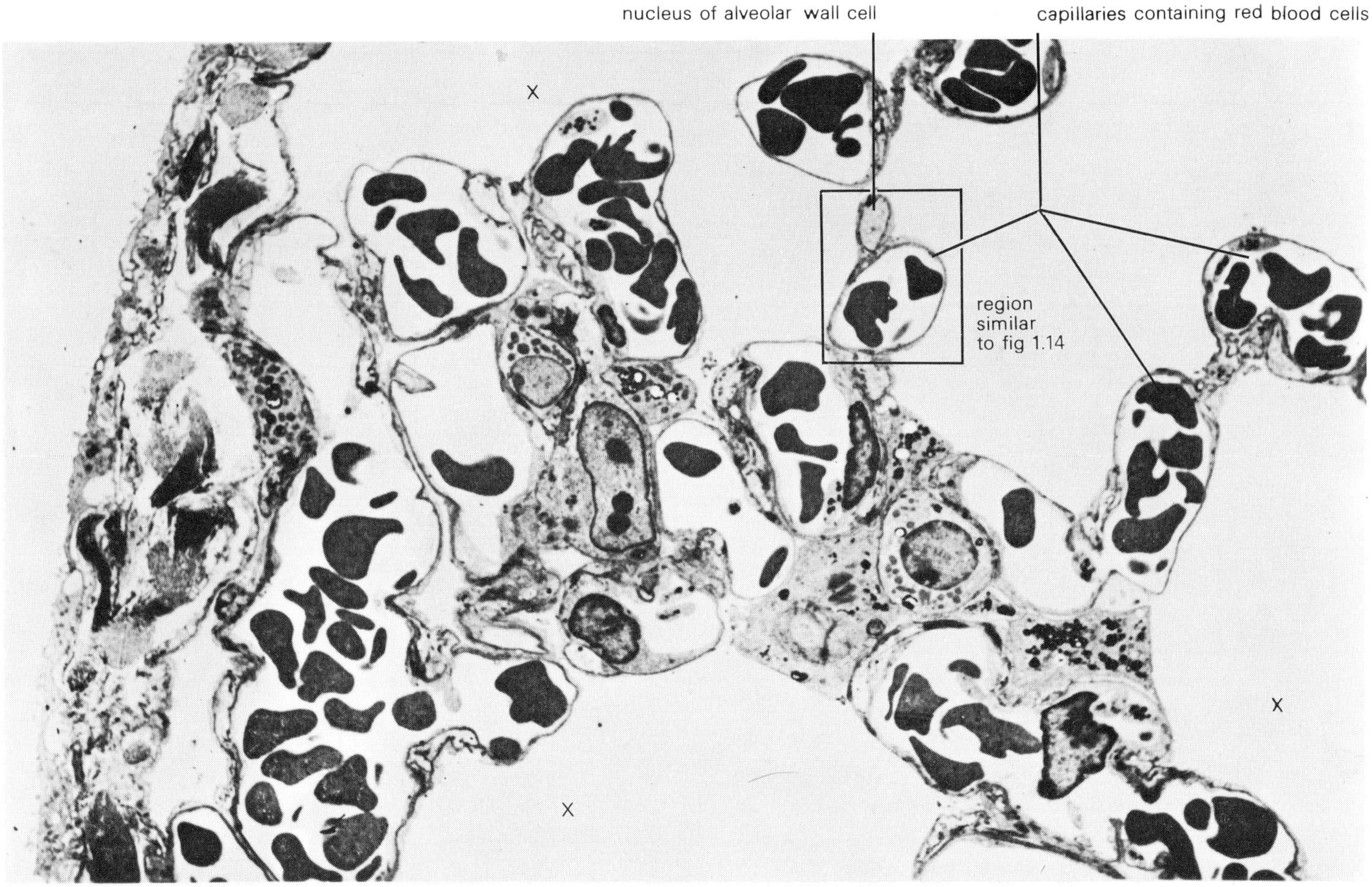

FIGURE 1·13

What does this show us about the way the capillaries lie in relation to the alveolar wall that is not shown in FIGURE 1.7 **?**

B FIGURE 1.13 **is an electron micrograph of a transverse section of part of a mouse lung. It looks very much the same as a portion of** FIGURE 1.12.

i What are the spaces marked x **?**

ii What may be said of the surface area of the capillaries relative to the spaces x **?**

C FIGURES 1.14a **and** 1.14b **are electron micrographs of three capillaries in a human lung, though they contain no alveolar wall cell nuclei. Taking** FIGURE 1.14b, **describe the relationship of the cytoplasm of the alveolar wall cell to that of the capillary wall cell.**

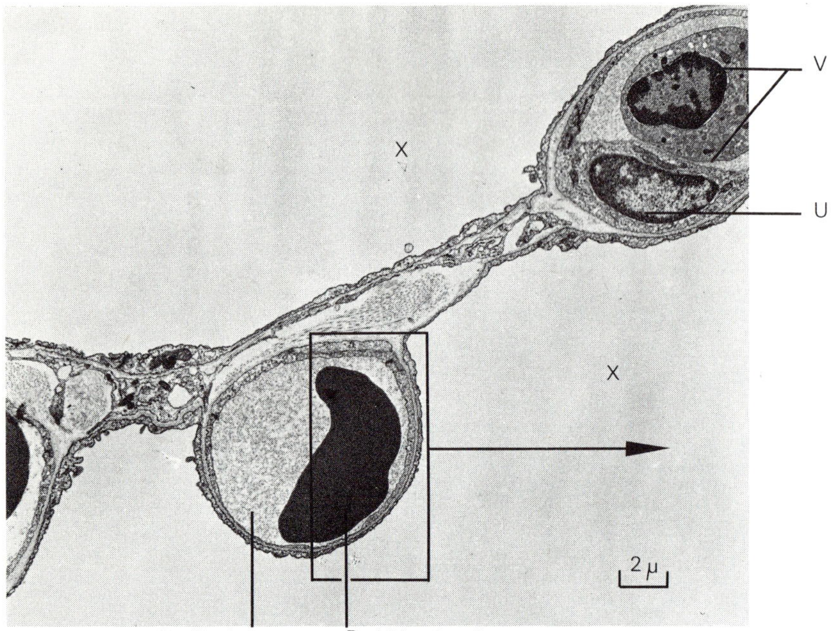

FIGURE 1·14a

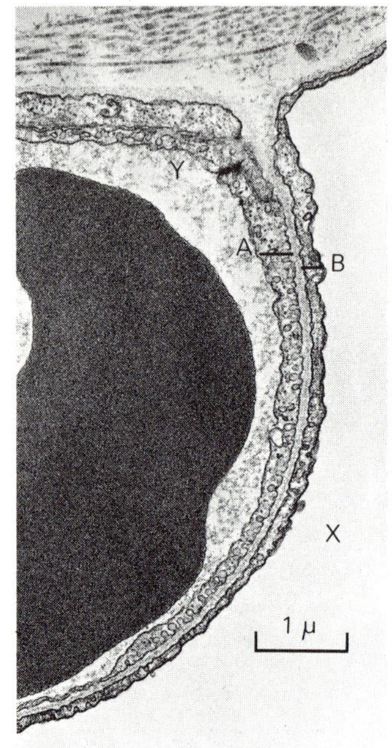

FIGURE 1·14b

14

D **FIGURES 1.14a and 1.14b are similar to the area in FIGURE 1.13 which has been boxed off. What are the spaces marked X ?**

E **What substance is present in life in the area marked Y ?**

F **Look at FIGURE 1.14b and describe the path followed by an oxygen molecule passing from the alveolus into the red blood cell. Name each structure through which it passes.**

G **Measure the width of the alveolar cell at the point A and of the capillary cell at the point B. Calculate their true thicknesses.**

H **Look at the bromine experiment shown in FIGURE 5.11 on page 74. This model can easily be fitted into FIGURE 1.14b. If we are thinking of the passage of oxygen from the alveolus into the red blood cell, which area of FIGURE 1.14b would correspond to the bottom gas jar and which to the top of the upper gas jar?**

I **What do you suppose the structures U and V in FIGURE 1.14a to be? Give the reasons for your answer.**

FIGURE 1.12 shows hair-like structures on the outside of the alveoli, labelled 'elastic fibres'. They form a basket-like mesh round the alveoli and can be seen better in FIGURE 1.15.

As the name suggests, they are very elastic and make the whole lung elastic—it has a strong tendency to shrink, even at the end of expiration.

FIGURE 1.16 shows two microscope slides with a film of water between them. Considerable force has to be exerted on the corks to pull the slides apart. If there is no water between the slides they come apart easily. The water molecules stick strongly to the glass and even more strongly to each other, making the wetted slides difficult to separate.

If you have not seen this demonstration, do it for yourself.

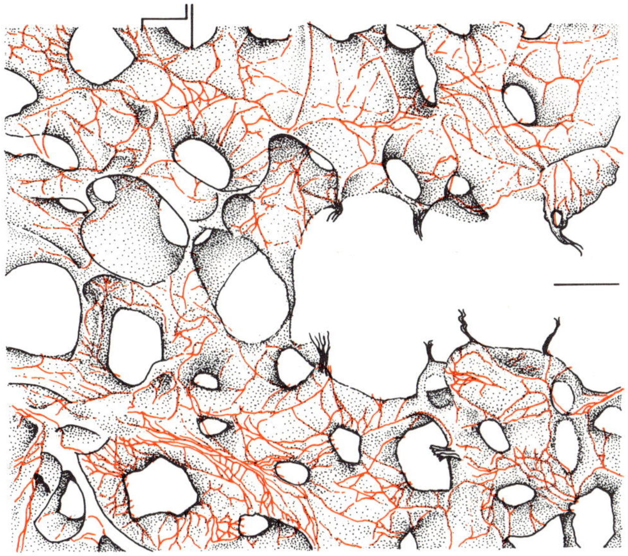

FIGURE 1·15

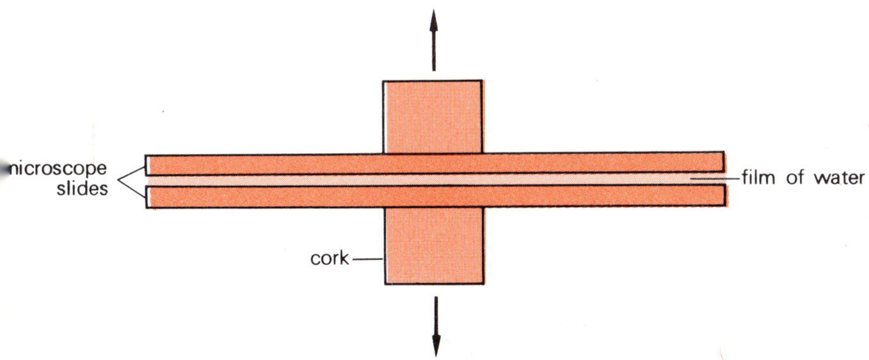

FIGURE 1·16

FIGURE 1.17 shows how FIGURE 1.16 can be used as a model to help understand the working of the chest.

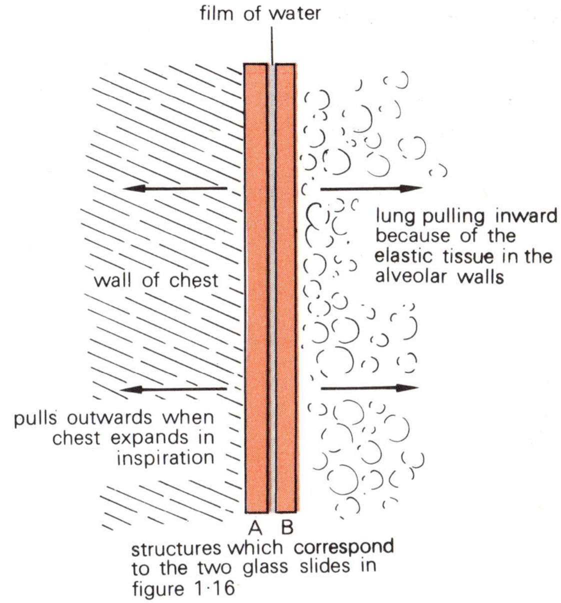

FIGURE 1·17

A **What are the structures A and B, which correspond to the two glass slides in FIGURE 1.16 ?**

B **What is the film of liquid corresponding to the water between the two slides?**

C **Explain how the structures A and B and the fluid between them contribute to the lung's ability to expand when the volume of the chest increases during inspiration.**

D **Explain the contribution made by these structures to the lowering of the rib cage and raising of the diaphragm during expiration.**

E **If air is allowed to enter the space between the lung and the wall of the chest (e.g. in an accident, or if a person is shot in the chest), the lung 'collapses', that is, it shrinks to a small size and is functionless.**

i **Explain why the lung shrinks.**

ii **Explain why it is functionless in this condition.**

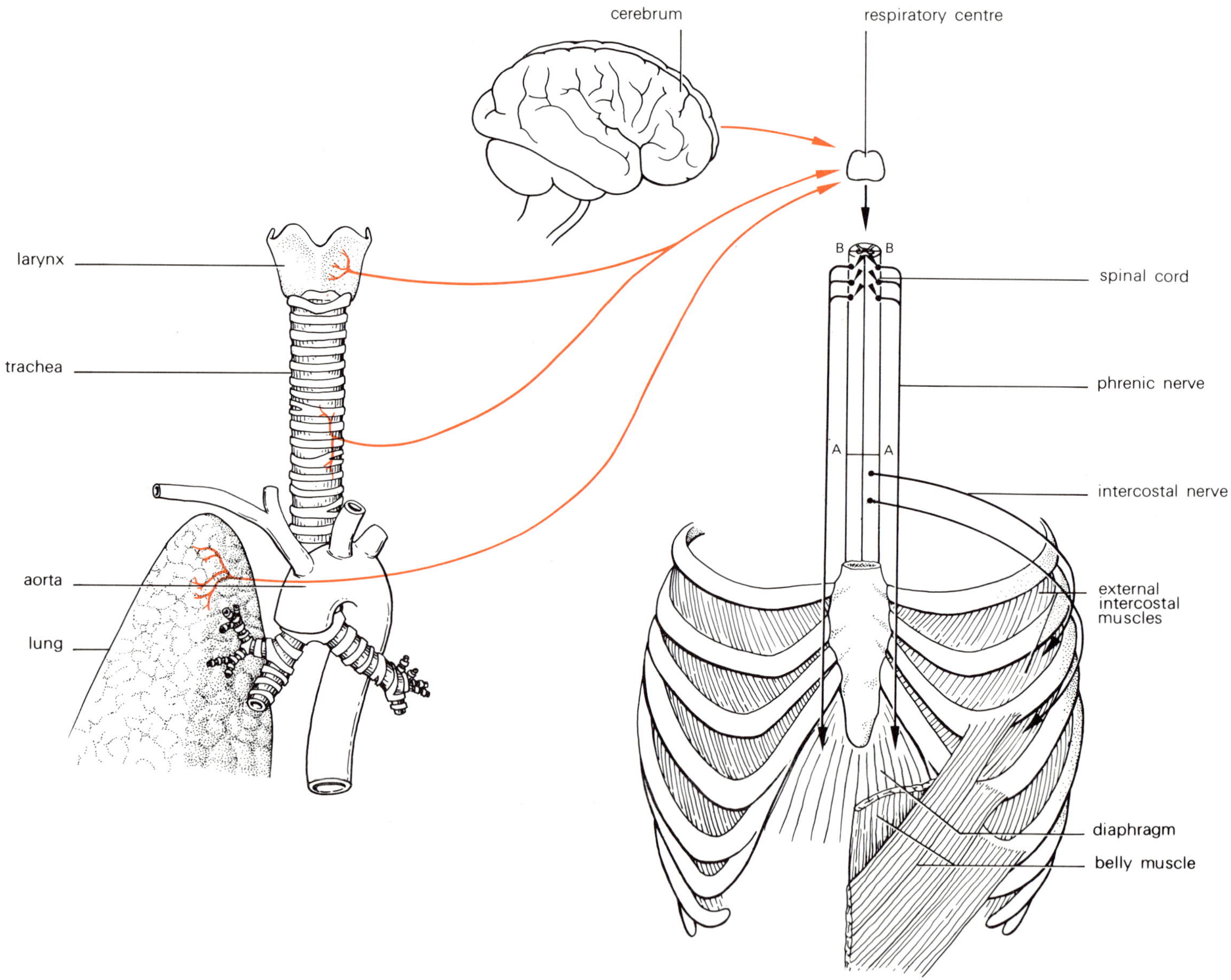
cerebrum
respiratory centre
larynx
trachea
aorta
lung
B
B
spinal cord
phrenic nerve
A
A
intercostal nerve
external intercostal muscles
diaphragm
belly muscle

FIGURE 1·18

15

We have seen in question 10**Eii** that the volume of the alveoli is 17 times greater than the volume of the dead space. The total surface area of the alveoli is very great. More important is the surface area of the capillaries exposed to the air in the alveoli (see FIGURES 1.2, 1.3 and 1.14): this has been calculated to be about 100 square metres.

We have also seen from FIGURES 1.14a and 1.14b that the wall of the alveolus and of the capillary against which it is pressed are very thin.

A Explain why these two facts are of great significance from the point of view of the rapid uptake of a large amount of oxygen in the lungs.

B How is the structure of the lobule related to the efficient use of lung space?

16

Breathing can to some extent be controlled voluntarily. For example, breathing rate can be modified during speaking or singing; and it is possible to hold one's breath for about 45 seconds. But breathing is essentially *involuntary*; it proceeds constantly from birth to death in wakefulness and sleep without our thinking about it. The rate and depth of breathing increase automatically when the body is more active, for example during running.

We know that inspiration is due to the contraction of the external intercostal muscles and the diaphragm, and that expiration is largely passive–when the external intercostals and the diaphragm relax, they return to the position they occupied before inspiration started.

The external intercostals and the diaphragm must, of course, contract in time with each other. The stimulus causing them to contract is conveyed along the *intercostal nerves* (supplying the external intercostal muscles) and the *phrenic nerves* (supplying the diaphragm).

Find these nerves in FIGURE 1.18.

You will see that the phrenic nerves arise from the spinal cord in the neck region and that the intercostal nerves arise from the spinal cord in the chest region. The cell bodies of these nerves are indicated by black dots in FIGURE 1.18, and their axons by black lines.

A If the spinal cord is cut across in the position marked A-A, above the origin of the intercostal nerves, the external intercostal muscles are paralysed, though neither the cell body nor the axon of the intercostal nerves have been harmed. The diaphragm continues to move.

What does this tell us about the nervous control of the external intercostal muscles and the diaphragm?

B If the spinal cord is cut across in the position marked B-B, above the origin of the phrenic nerves, both the diaphragm and the external intercostal muscles are paralysed.

What does this tell us about the nervous control of the diaphragm and the external intercostal muscles?

C FIGURE 1.19 **is a drawing of a longitudinal section through the centre of a cat's brain. When cuts are made across the brain in the positions W-W, X-X and Y-Y, that is behind the thalamus, in front of the cerebellum and at the front of the medulla oblongata, breathing continues. When the brain is cut across in the position Z-Z, at the rear of the medulla oblongata, breathing ceases.**

i What does this evidence tell us about the location in the brain of the control centre for respiration?

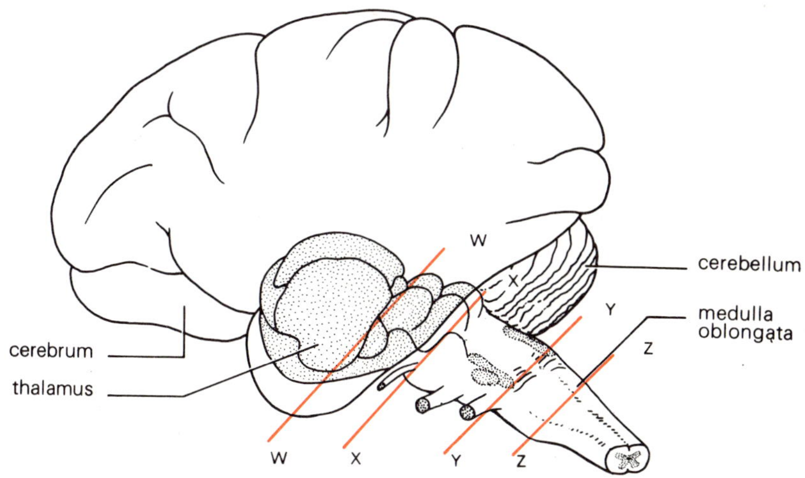

FIGURE 1·19

ii Taking this evidence with that put forward in questions A and B above, what deduction can we make about the nervous connection between the brain control centre for respiration and the cell bodies of the phrenic and intercostal nerves?

17

The part of the brain controlling respiration is called the *respiratory centre*. You can see from FIGURE 1.18 that it is influenced by nerves from many parts of the body–from the cerebrum (the 'thinking' part of the brain; these influences enable breathing to be controlled to some extent voluntarily, as we have seen); from the lung; from the larynx and trachea; and from other places not shown on this diagram.

All these influences affect the rate of breathing by affecting the respiratory centre, from which pass, down the spinal cord to the cell bodies of the phrenic and intercostal nerves, the impulses causing muscular contractions.

The respiratory centre is also influenced by the slightest rise in the carbon dioxide content of the blood. If the body becomes active, as in running, more carbon dioxide is produced in the muscle cells. The carbon dioxide concentration in the blood rises. More carbon dioxide is released into the alveoli from the blood. The normal concentration of carbon dioxide in the alveolar air is about 5·6%: a rise of only 0·2% causes a change in breathing which doubles the volume of air breathed per minute. Thus the excess carbon dioxide in the alveoli is blown out of the lungs and more oxygen is taken into the lungs.

The effect of carbon dioxide on respiration can be demonstrated by allowing a person to breathe air which contains more than the normal concentration, which is 0·04%. Increasing the carbon dioxide content of alveolar air will cause a rise in the carbon dioxide concentration in the blood; when this blood circulates through the respiratory centre it will affect the nerve cells of the centre, causing them to increase the rate and depth of respiration.

The results of such an experiment are shown in FIGURE 1.20.

A Increasing the carbon dioxide concentration of the air breathed in from normal (0·04%) to 3·07% has almost doubled the depth of each breath (lines 1 and 2).

1	Percentage of CO_2 in inspired air	0·04	0·79	2·02	3·07	5·14	6·02
2	Average depth of respiration, cm^3	673	739	864	1216	1771	2104
3	Number of breaths per minute	14	14	15	15	19	27
4	Amount of air drawn into the alveoli in unit time (100 = normal)	100	116	153	226	498	857
5	Percentage of CO_2 in alveolar air	5·6	5·5	5·6	5·5	6·2	6·6

FIGURE 1·20

i What effect has this increase in depth of breathing had on the carbon dioxide concentration of the alveolar air (line 5, note that the normal carbon dioxide concentration of alveolar air is 5·6%)?

ii Suppose the depth of breathing did not change when the carbon dioxide concentration of the air breathed in increased. What then would you expect to be the concentration of carbon dioxide in the alveolar air when the concentration in the inspired air was 3·07%?

iii Line 4 shows how deeper breathing affects the amount of air entering the alveoli in unit time. How do these data explain the percentage carbon dioxide values of alveolar air seen in line 5 of FIGURE 1.20, up to the value 3·07% carbon dioxide in the air breathed in?

B What is the effect on breathing rate of increasing the carbon dioxide content of inspired air from 0·04% to 3·07%?

C How many more breaths are taken per minute when the inspired air contains 5·14% carbon dioxide?

D How much air is taken in per minute

i under normal conditions?

ii when the inspired air contains 6·02% carbon dioxide?

18

FIGURE 1.21 shows in line A the effect of increased oxygen pressure in the air in contact with a solution of haemoglobin (x axis), on the amount of oxygen which the solution takes up. Thus when the oxygen pressure is 10 mm mercury, about 57% of the haemoglobin is in the form of oxyhaemoglobin.

In the case of mammalian blood, the haemoglobin is present inside red blood cells and it does not then behave in the same way as a solution of haemoglobin. The percentage of oxygen absorbed by blood when exposed to varying external pressures of oxygen is shown in line B.

This change of behaviour is of great advantage in the transport of oxygen to the respiring tissues.

A **The atmospheric pressure in the alveoli** due to the oxygen fraction of the air **is about 100 mm mercury. The oxygen pressure in blood leaving exercising muscle may be as little as 10 mm Hg.**

i **What percentage of oxygen is given up by blood when it passes from a region where the external oxygen pressure is 100 mm Hg to one where the external oxygen pressure is 10 mm Hg?**

ii **What percentage of oxygen is given up by a solution of haemoglobin when the external oxygen pressure changes from 100 mm Hg to 10 mm Hg?**

B **Oxygenated blood travels from the heart along the aorta, arteries and arterioles to the capillaries. In the walls of the arteries the oxygen pressure is about 60 mm Hg. In the tissues surrounding the capillaries the oxygen pressure may be as little as 10 mm Hg.**

From curve B **in** FIGURE 1.21, **what percentage of its oxygen content will the blood give up:**

i **in passing along the arteries**

ii **in passing through the capillaries?**

C **How does your answer to B show that blood is an efficient carrier of oxygen to the respiring tissues which need it?**

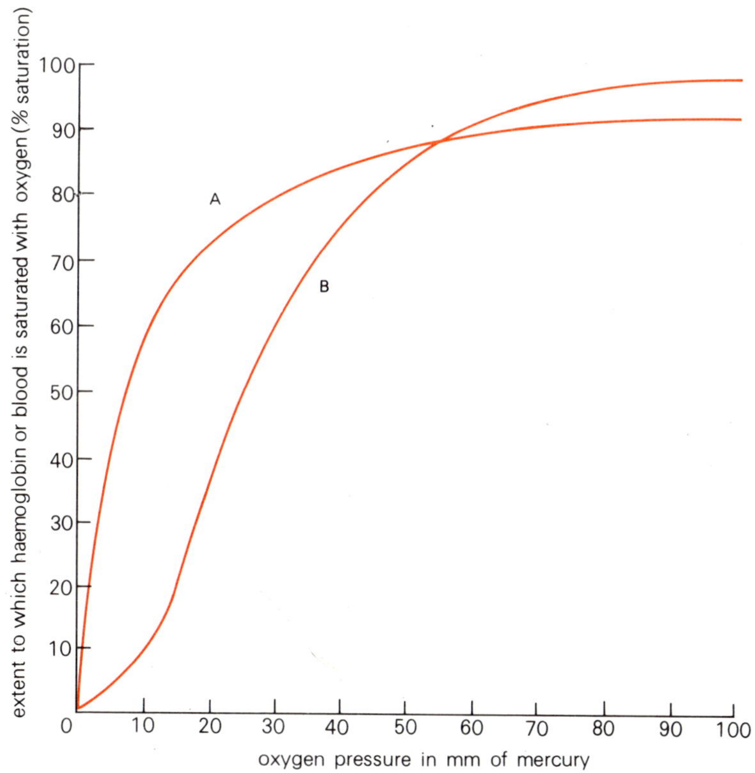

FIGURE 1·21

19

The shape of the curve shown in FIGURE 1.21 is influenced not only by the oxygen pressure in the medium (be it air or living tissue) surrounding the blood, but also by the pressure of carbon dioxide in this medium.

The curve in FIGURE 1.21 was obtained when the carbon dioxide pressure in the medium surrounding the blood was 40 mm Hg.

FIGURE 1.22 shows us the curves obtained when the carbon dioxide pressure in the medium surrounding the blood was 0, 3, 20, 40 and 90 mm Hg.

Take, for example, the point on the x axis where the pressure of oxygen in the tissue surrounding the blood is 10 mm Hg (the condition for actively respiring muscle).

A **To what extent is the blood saturated with oxygen when the carbon dioxide pressure in the medium surrounding it is 90 mm Hg, as it would be in respiring muscle?**

B **To what extent is the blood saturated with oxygen when the carbon dioxide pressure in the medium surrounding it is 0 mm Hg?**

C **How, then, does the pressure of carbon dioxide in the medium surrounding the blood affect the affinity of blood for oxygen?**

D **How does your answer to C show that because of this effect of carbon dioxide the blood is more efficient in its job of picking up oxygen at the lung and releasing it in the respiring tissue?**

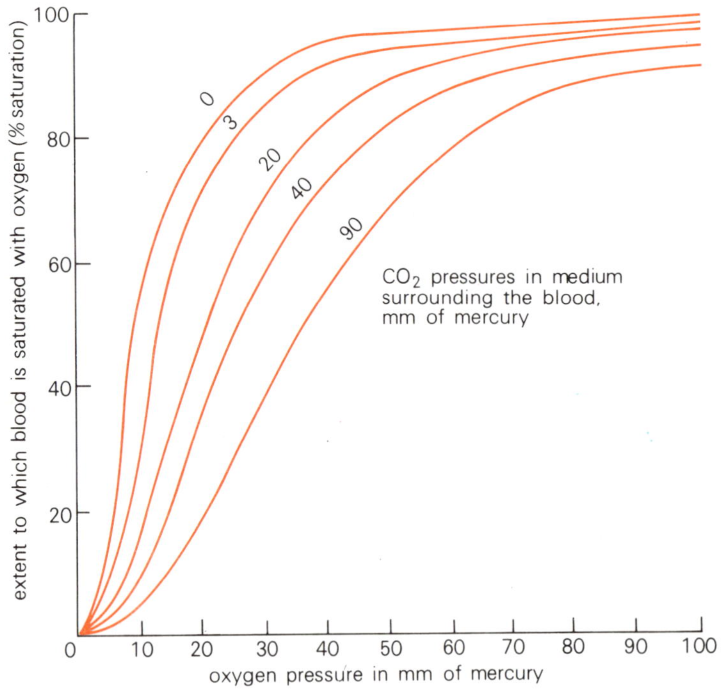

FIGURE 1·22

20

The word 'respiration' can be used to mean:

i breathing

ii the exchange of oxygen and carbon dioxide between living cells and the atmosphere.

State a third meaning, involving glucose, oxygen **and** energy.

21

Glucose burns well in air; it reacts with oxygen forming carbon dioxide and water vapour. Energy is given out in the forms of heat and light. Glucose is therefore a fuel: it could, for example, be used to raise steam and work a steam engine.

Glucose is also the fuel of living cells, but it does not burn in the cell in the form of a little fire.

A How does the oxidation of glucose in the living cell differ from its oxidation when burning in air

i from the point of view of the forms of energy released?

ii from the point of view of the suddenness with which the energy is released?

B How are the chemical substances ADP and ATP involved in gathering up the energy released in the oxidation of glucose in the living cell?

C ATP stands for adenosine tri-phosphate. It has a complicated chemical formula, but we can represent it simply as A – P ~ P ~ P.

If the link between the first and second or between the second and third P (which stands for phosphate) is broken, energy is set free. The same is true if the link between the adenine molecule (A) and the first P is broken, but this sets free only one quarter the amount of energy liberated when the second or the third link is broken. That is why the second and third are represented as ~ and the first as a dash –.

Explain the importance of the ~ link (the 'energy-rich' bond) in the utilization in the cell of the energy from the oxidation of glucose which has been 'captured' in the ATP molecule.

D One example of the use of energy from ATP is that it provides energy for the contraction of muscle.

i For what other purpose is energy from ATP used in the epithelial cells of the villi and in the tubule cells of the uriniferous tubules in the kidney?

ii What is the general use to which the energy from ATP is put in the living cell?

iii What is the name of the minute structures (organelles) within living cells where ATP is generated by the oxidation of glucose?

22

The process by which energy is liberated within cells by the oxidation of glucose is called *tissue respiration.* There are some twenty to thirty chemical steps between glucose and the ultimate products, carbon dioxide and water.

This *gradual* breakdown of glucose enables energy to be released in stages and to be gathered up by the conversion of ADP to ATP for use in all the cell processes which require energy.

The details of the chemical processes involved are beyond our scope, but the complex whole can be reduced to certain important essentials which are reasonably easy to understand.

Overall, the two most important things to appreciate are

i whereabouts in the chain of chemical reactions chemical energy is trapped in the form of ATP;

ii how the chain of chemical reactions is affected by the availability of oxygen or the lack of oxygen.

First, read the following description of the process. It is summarized in FIGURE 1.23a and b. When you can understand FIGURE 1.23a and b you will be in a position to answer the questions on it which follow.

Glucose and other sugars are absorbed in the small intestine and are carried in the blood to the liver and muscles, where they are converted into insoluble glycogen for storage (see questions on Bernard's experiments pp 76–77).

Glycogen means 'giving rise to glucose'. Glycogen is very similar to starch and is easily converted back to soluble glucose. This can then be carried in the blood to respiring tissues for use to produce energy, or to muscles to replenish glycogen which has been consumed there (FIGURE 1.23a).

We will follow the fate of glycogen in the muscle, though the oxidation of glucose in the other tissues is exactly the same process.

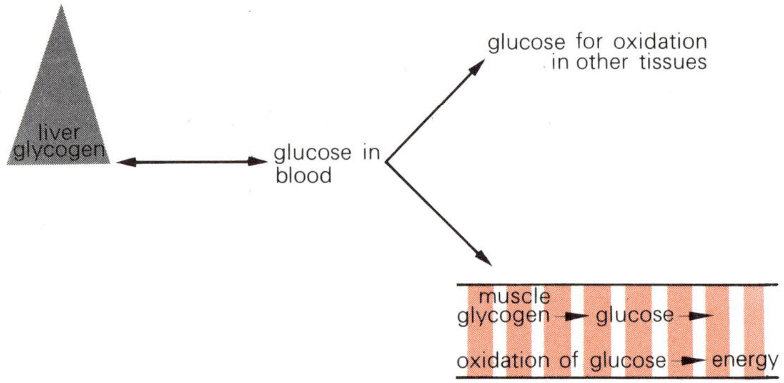

FIGURE 1·23a

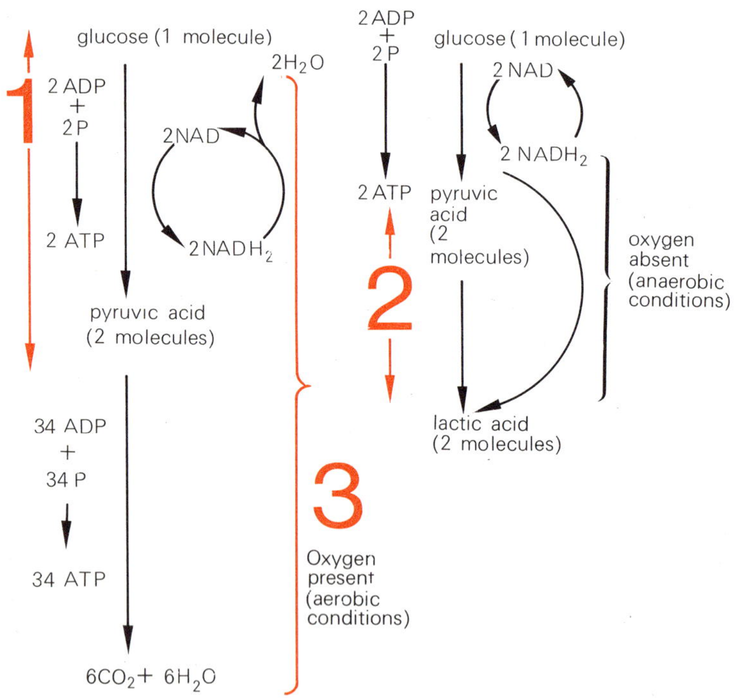

FIGURE 1·23b

The many chemical steps between glucose and the end products, carbon dioxide and water, can be divided into two stages.

Stage I Steps result in the conversion of one molecule of glucose, which contains six carbon atoms (formula $C_6H_{12}O_6$) into two molecules of *pyruvic acid* (formula $C_3H_4O_3$), each of which contains three carbon atoms.

Energy is liberated during this conversion. Two molecules of ADP are converted into two molecules of ATP by joining with two phosphate groups and absorbing energy, i.e. 2ADP + 2P + energy→2ATP.

Stage III Bypassing *stage II* in FIGURE 1.23b for a moment, we see that *stage III* steps result in the conversion of pyruvic acid to carbon dioxide and water; and the energy liberated results in the formation of 34 molecules of ATP from 34 molecules of ADP for each molecule of glucose used (i.e. 34ADP + 34P + energy→34ATP).

Thus we have seen whereabouts in the chain of chemical reactions chemical energy is trapped in the form of ATP.

The other point we had to look for was how the whole process is affected by the availability or lack of oxygen. *Stages I and III* proceed as we have described *when oxygen is freely available* (*aerobic* conditions).

Stage II Taking the other extreme, where there is *no oxygen available* (*anaerobic* conditions), *stage I* proceeds as before, that is glucose is converted to pyruvic acid; but *stage III* reactions do not take place. Instead, the pyruvic acid is converted into *lactic acid*, formula $C_3H_6O_3$. The whole of *stage III* is omitted: instead of pyruvic acid becoming oxidized to carbon dioxide and water, it is all converted into lactic acid.

The reason is shown on the right hand side of FIGURE 1.23b. To understand it we must know about one more detail of the chemical changes in *stage I*.

An essential stage in the conversion of one molecule of glucose into two molecules of pyruvic acid is that a substance which is present in only very small quantities in the cell (a *coenzyme* called NAD), takes two of the hydrogen atoms from the glucose molecule, becoming $NADH_2$. More glucose cannot be converted to pyruvic acid unless more NAD is available. If oxygen is present, the hydrogen from $NADH_2$ is removed and passed on to oxygen, forming water: thus NAD is restored, becomes $NADH_2$ again, and so on.

If oxygen is *not* available one might suppose that the conversion of glucose to pyruvic acid would come to a quick halt, once all the NAD had been locked up in the form of $NADH_2$.

However, $NADH_2$ can react with *pyruvic acid* forming *lactic acid* and restoring NAD. (Note that the formulae of pyruvic acid ($C_3H_4O_3$) and lactic acid ($C_3H_6O_3$) differ only by two hydrogen atoms.)

Thus more glucose is converted to pyruvic acid, forming more $NADH_2$, which reacts with pyruvic acid forming lactic acid and restoring NAD, and so on.

A i What is the name of the blood vessel which connects the small intestine to the liver, in which freshly-absorbed glucose passes from the gut to the liver?

ii What is the name of the blood vessels leaving the liver, taking glucose from the liver into the general circulation? Which vessel do they join soon after leaving the liver?

B What hormones do you know of which affect the storage of glycogen in the liver? What effects do they have on the storage of glycogen?

C From FIGURE 1.23b, **when does the major part of the energy release during the oxidation of glucose take place? Say how** FIGURE 1.23b **tells you this.**

D *Stage I* **in** FIGURE 1.23b **can take place in the cell whether oxygen is available or not available. If oxygen is not available, the energy-releasing stages end here and lactic acid accumulates. How, then, does such** anaerobic **respiration (i.e. the breakdown of glucose to pyruvic acid in the absence of oxygen) compare with** aerobic **respiration with regard to its yield of chemical energy?**

E The lactic acid formed in muscle under anaerobic conditions is not wasted. The liver is able to convert it back to glycogen.

i How would the lactic acid from within the muscle cell reach the liver?

ii Working within FIGURE 1.23b **and the description which followed it, make a guess at a possible fate for lactic acid when oxygen becomes available after a period of oxygen lack.**

23

We will now examine *stage II* more carefully.

If no oxygen is available (anaerobic conditions), lactic acid is formed from pyruvic acid; this enables NAD to be formed from $NADH_2$.

We might think, since blood carrying oxygen is flowing through the capillaries of the muscles, that lactic acid would never be formed. Instead pyruvic acid would be oxidized to carbon dioxide and water, generating 34 molecules of ATP.

The body has a store of fuel for respiration in the form of glycogen. It has no corresponding store of oxygen. The six litres of blood in the body can hold about one litre of oxygen.

FIGURE 1.24, COLUMN 2 shows how fast oxygen is used when the body is performing various tasks.

EXERCISE	1 VOLUME OF BLOOD PUMPED OUT OF EACH VENTRICLE (litres/min)	2 TOTAL AMOUNT OF OXYGEN USED BY THE BODY (litres/min)
Rest	5	0·25
Walking 3·2 km/h	10	0·8
Walking 8 km/h	20	2·5
Running 12 km/h	25	3·0
	34	4·0

FIGURE 1·24

A How long would the one litre of oxygen held in the blood last if it were not replenished when the body was

i resting?

ii running at 12 km/h?

B Obviously the body responds to increased demand for oxygen by obtaining more of it.

i What changes in inspiration cause more air to be taken into the lungs during exercise?

ii What stimulus causes the changes noted in i ?

C FIGURE 1.24 **shows that the amount of blood pumped out of each ventricle increases as exercise becomes more vigorous.**

There are two ways in which the heart can increase its output of blood per unit of time. What are they?

D TABLE 1.24 **shows that between rest and walking at 3·2 km/h the volume of blood given out by each ventricle per minute has doubled. Between rest and walking at 8 km/h**

it has quadrupled; between rest and running at 12 km/h it has multiplied five times; between rest and very severe exercise it has multiplied about seven times.

How does the increase in the total amount of oxygen used by the body over this range of activities compare with these increases?

E **The faster blood flows through the lung the more oxygen it will pick up in unit time; but as we see from D this increase in blood flow does not fully account for the rise in oxygen used and therefore taken up.**

Another factor here is that as activity increases, so the blood arriving at the lung has a greater affinity for oxygen. By referring to FIGURE 1.22, **explain why this is so.**

24

Thus although the body has no store of oxygen, it responds to increased demand for oxygen during exercise by increasing the rate and depth of respiration and increasing the rate of blood flow through the lungs and to the respiring muscles.

We must ask: is this enough to satisfy *stage III* in FIGURE 1.23b, that is, is the oxygen supply enough to oxidize the pyruvic acid produced in *stage I*, or will some or all of this pyruvic acid have to be converted to lactic acid?

FIGURE 1.25 gives us an immediate answer. It shows the changes in concentration of lactic acid in the blood before, during and after a short period of vigorous exercise.

A **What is the concentration of lactic acid in the blood before exercise (the 'resting level')?**

B **How does this concentration change**

i **during the period of exercise (one minute)?**

ii **in the nine minutes immediately following the period of exercise?**

C **If lactic acid is produced in muscle** during **exercise, how do you account for the difference between your answers to i and ii in question B ?**

D **What does this graph tell us about the body's ability to obtain enough oxygen during exercise to oxidize the pyruvic acid produced in** *stage I* **of** FIGURE 1.23b **?**

E **How is the glucose providing the energy for muscle contraction in these conditions (i.e. in vigorous exercise)?**

F **What chemical reaction is giving rise to lactic acid in these conditions (of exercise), and why is it important that it should take place?**

25

Thus when muscles start to exercise, more oxygen is needed than can be supplied. The muscles derive their energy from the con-

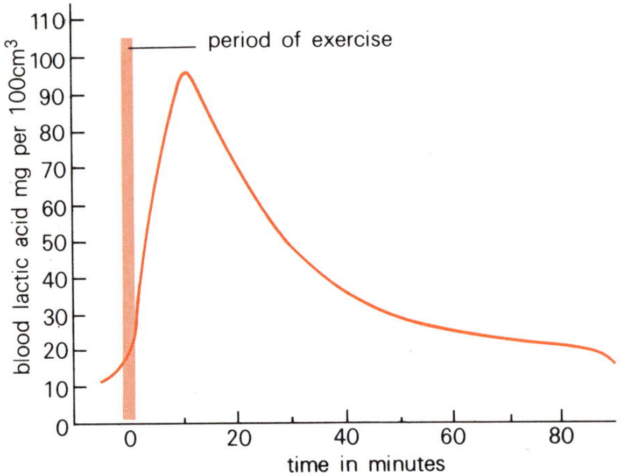

FIGURE 1·25

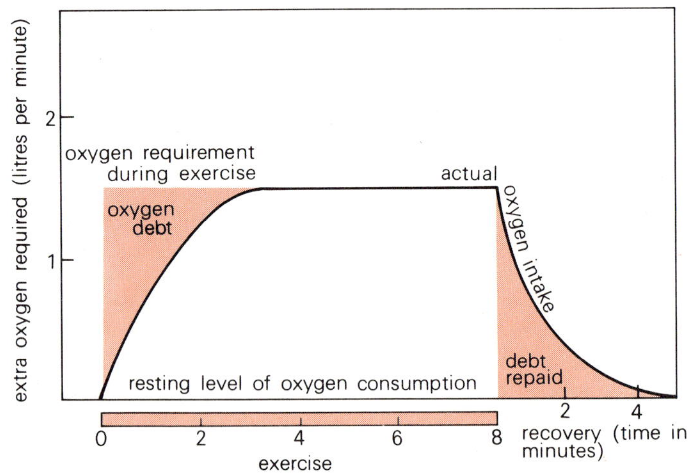

FIGURE 1·26

version of glucose to pyruvic acid, which produces lactic acid.

When exercise finishes, some of the lactic acid formed is converted into glycogen by the liver. The remainder is converted back to pyruvic acid, which is then oxidized by the extra oxygen taken in after exercise has finished. The extra oxygen comes from frequent and deep breathing which continues after the exercise, as anyone who has run a race knows well. Indeed it is possible for an athlete to run 100 m holding his breath and taking in no oxygen until after the race is run.

It is said that in producing energy by *stage I* and *II* reactions, the body has incurred an *oxygen debt* that has to be 'repaid' after the exercise has finished.

FIGURE 1.26 shows how an oxygen debt has been incurred during a mild spell of exercise such as walking at 8 km/hr.

A **From FIGURE 1.26, what is the extra oxygen requirement immediately the exercise starts?**

B **How long is it before the intake of oxygen equals this extra requirement?**

C **During this time an oxygen debt is incurred (i.e. energy is derived from the reaction glucose→pyruvic acid→lactic acid), which must be repaid by the subsequent oxidation of the lactic acid. From FIGURE 1.26, when is the debt repaid, and how long does it take to repay it?**

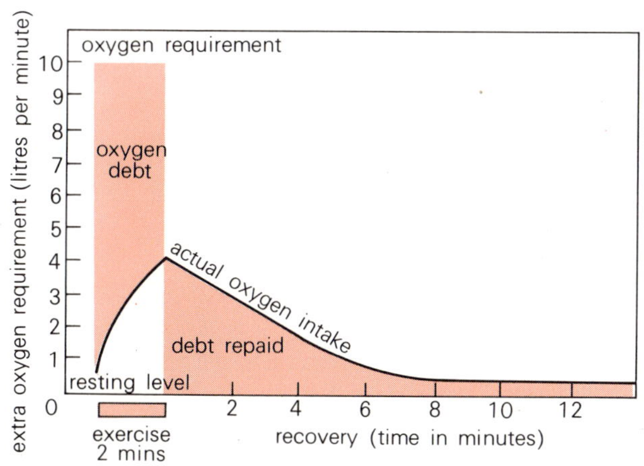

FIGURE 1·27

D **No further oxygen debt is incurred after four minutes of exercise. The oxygen requirement is then equalled by the extra oxygen taken in. What does this tell you about the pyruvic acid that is formed in the muscles after four minutes of exercise?**

E **The oxygen debt is not paid off until after the exercise has ceased. If we did not know that this was a fact, what might we have guessed would happen to the lactic acid formed in the first four minutes of exercise?**

26

FIGURE 1.27 shows the oxygen debt incurred during a short period (two minutes) of vigorous exercise.

A **What was the extra oxygen requirement immediately exercise started?**

B **What was the highest level of extra oxygen intake attained during the period of exercise?**

C **What, then, at the end of the period of exercise, was the difference between oxygen needed and oxygen taken in (in litres per minute)?**

D **How long did it take to repay this oxygen debt (by deep and frequent breathing after exercise finished)?**

E **Comparing FIGURE 1.27 with 1.26, what is the difference between the balance of oxygen needed and oxygen intake at the end of a period of mild exercise and at the end of a period of severe exercise?**

F **What does the answer to E tell you about the body's ability to cope with the demand for extra oxygen during exercise?**

2 Blood and the circulation

27

FIGURE 2.1 shows in line A the influence of oxygen concentration outside human red blood cells on the percentage of haemoglobin present as oxyhaemoglobin within the cells.

A **The portion of atmospheric pressure due to oxygen is about 150 mm Hg, but the pressure of oxygen in the alveoli is only about 100 mm Hg. Remember the difference in the size of the chest during inspiration and after expiration.**

How can you account for the fact that the pressure in the lung due to oxygen is less than the pressure in the air immediately before breathing in?

B **In the blood leaving exercising muscle, the oxygen pressure may be as little as 10 mm Hg.**

i **From** FIGURE 2.1, **what percentage of oxygen has been given up by the oxyhaemoglobin in the blood from the time it has left the lung to the time when it is leaving the the muscle?**

ii **In birds there is a clear distribution of muscles which move the skeleton into two types, 'red' and 'white', giving the familiar red and white meats. The red muscle cells contain a red pigment called** myoglobin.

In mammals there is sometimes a similar distinction, as in rabbits; but in man, although both types of muscle cell exist, there is no sharp colour distinction. Red and white muscle cells are mixed. It seems that muscles concerned with movement (such as the biceps) contain a large proportion of white muscle cells. Those concerned with holding the body upright and with holding bones firm while other muscles contract against them (e.g. the trapezius muscle which holds the shoulder blade still) contain a large proportion of red muscle cells.

The function of myoglobin in red muscle is not known, but it has been suggested that it might act as an oxygen reservoir.

Suppose myoglobin were the respiratory pigment of red blood cells instead of haemoglobin. Consider the same situation as in question Bi, a pressure due to oxygen of 100 mm of mercury at the alveoli and of 10 mm of mercury in the blood leaving the exercising muscle.

Explain from line B **of** FIGURE 2.1 **why myoglobin would not be nearly as efficient as the haemoglobin in the blood is.**

28

FIGURE 2.2 shows the blood pressure in various parts of the human circulation. The pressure on the blood entering the aorta at the end of the contraction of the left ventricle is between 120 and 130 mm Hg (point X on the figure). The pressure on the blood in the aorta *before* the ventricle contracts is about 75 mm Hg (point Y on the figure). This is due to the squeezing action of the elastic wall of the aorta on the blood it contains.

A **What is the pressure exerted on the blood in the left ventricle when it has relaxed (point** Z **on** FIGURE 2.2**)?**

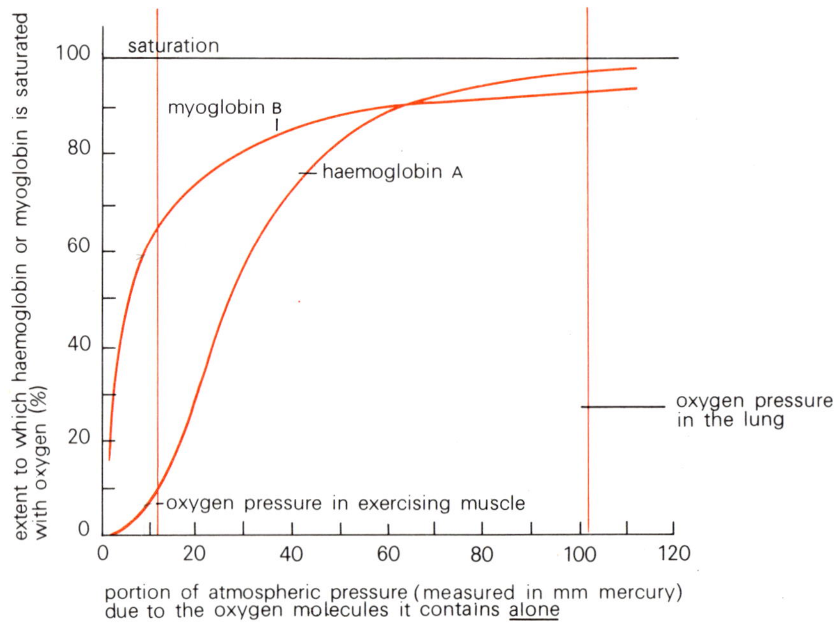

FIGURE 2·1

B The pressure in the aorta is not constant but fluctuates.

- **i** Why is this so?
- **ii** What is the maximum pressure in the aorta and when is it reached?
- **iii** What is the minimum pressure in the aorta, and when is it reached? What is causing the pressure on the blood at this time?
- **iv** What is the average pressure in the aorta?

C What is the pressure on the blood in the larger veins, just before it enters the right auricle?

D What, therefore, is the pressure difference between blood in the aorta (average value) and blood in these veins, that is the pressure which drives blood along the circulation?

E

- **i** What is the pressure on the blood in the pulmonary artery, just before the right ventricle contracts (the pressure due to the squeezing action of the pulmonary artery on the blood it contains)?
- **ii** What is the maximum pressure exerted on the blood by contraction of the right ventricle?
- **iii** What, then, is the average pressure on the blood in the pulmonary artery?
- **iv** Find the pressure on the blood in the pulmonary vein just before it enters the left auricle.
- **v** What, therefore, is the pressure difference between blood in the pulmonary artery (average value) and blood in the pulmonary veins, that is, the pressure which drives blood through the lungs?

F At which point on FIGURE 2.2 will the aorta be

- **i** at its widest
- **ii** at its narrowest?

Explain your answer.

G From FIGURE 2.2, in which part of the circulation from left ventricle to right auricle is there the greatest resistance to the flow of blood? Explain how you arrive at your answer.

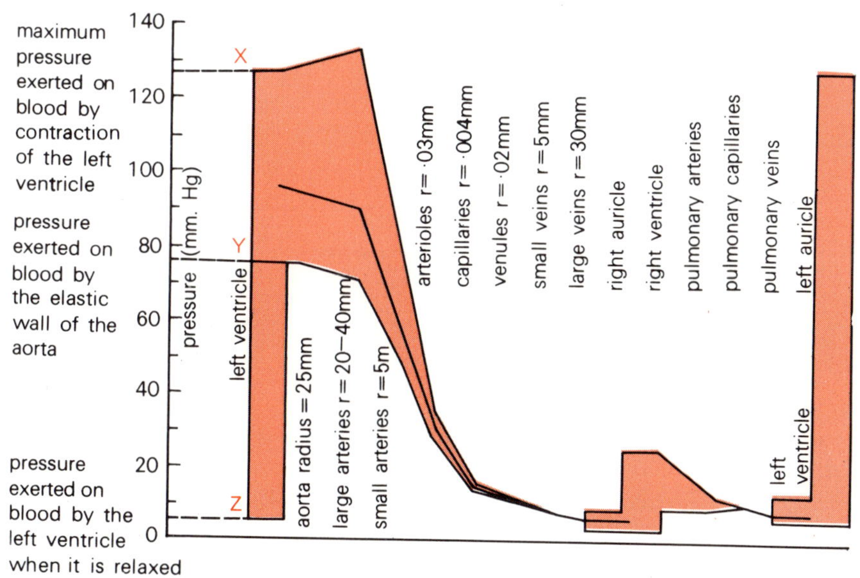

FIGURE 2·2

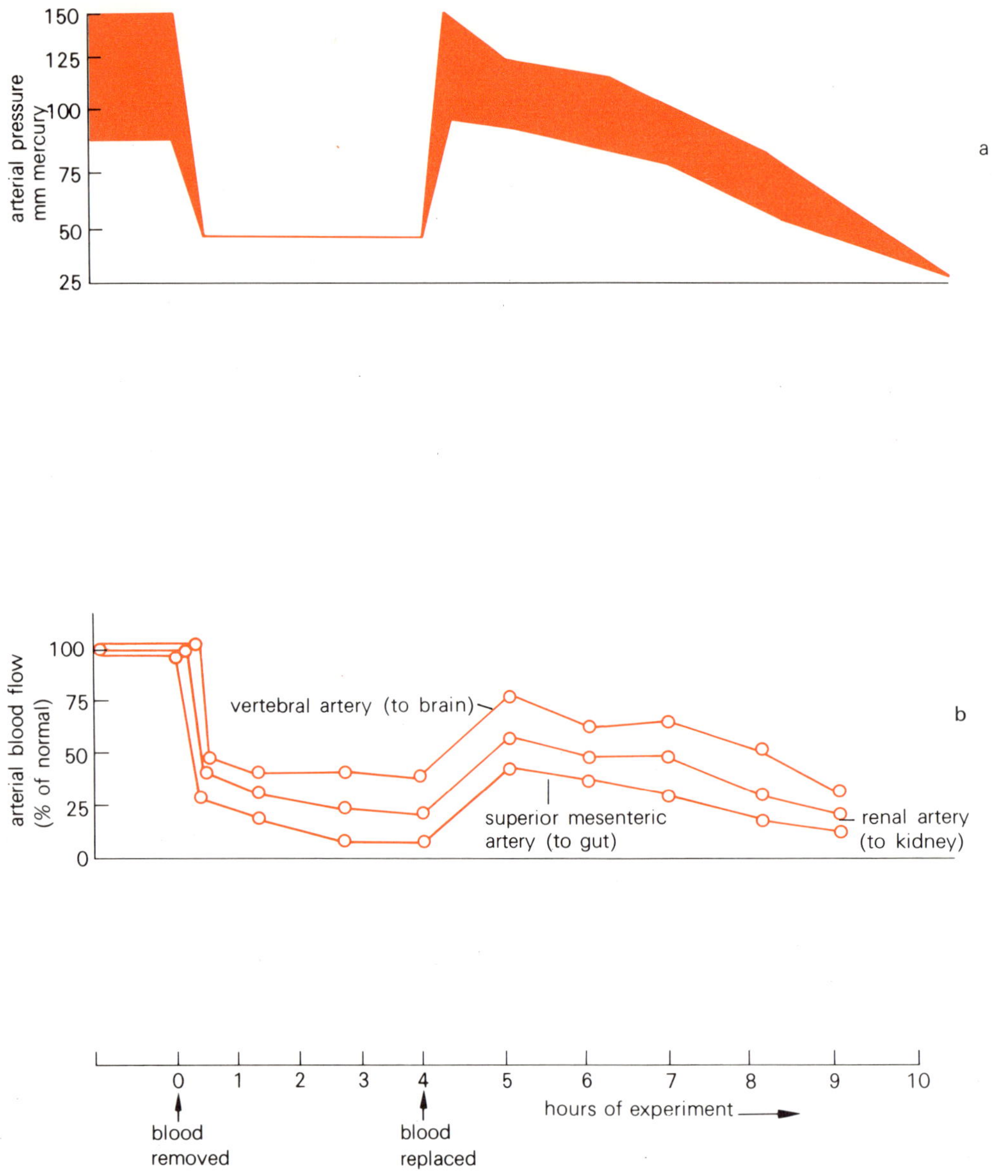
150
125
100
75
50
25
arterial pressure
mm mercury
a
100
75
50
25
0
arterial blood flow
(% of normal)
vertebral artery (to brain)
superior mesenteric
artery (to gut)
renal artery
(to kidney)
b
0
1
2
3
4
5
6
7
8
9
10
hours of experiment
blood
removed
blood
replaced

FIGURE 2.3

29

FIGURE 2.3a shows some of the results of an experiment in which half of the total volume of blood in a dog was removed and replaced four hours later.

The X axis shows the timing of the experiment.

A What was the average pressure on the blood in the aorta before the blood was removed?

B What happened to the pressure of blood in the arteries:

i when the blood was removed

ii for the next four hours

iii when the blood was replaced?

30

FIGURE 2.3b shows the changes in the flow of blood to three parts of the body, the brain, the kidney and the gut during the course of the experiment. The figures on the Y axis are percentages of normal flow, so at the start of the experiment the flow to all these parts was 100% normal.

A Describe the changes in the flow of blood to the brain, kidney and gut before bleeding, after loss of blood and during the hour following replacement of blood.

B Find the lowest level to which the blood flow dropped in each of these organs.

C The pattern of change in blood flow during the experiment is the same for all three organs, but the actual levels are different for each organ. During the period between removal and replacement of the blood, the animal had only half of its normal volume of blood.

How does FIGURE 2.3b tell you that this reduced amount of blood was put to the best possible use in keeping the animal alive?

31

The rate of heartbeat in an adult man sitting at rest is about 70 to 80 per minute; each beat therefore lasts $\frac{60}{70}$ seconds or about 0·8 s. Both auricles contract together (auricular systole), forcing blood into the ventricles. Both auricles and both ventricles rest for a short time (period of diastasis); then both ventricles contract together (ventricular systole).

FIGURE 2.4 shows these events drawn against a time scale when the heart is beating 75 times a minute. (Auricular and ventricular *diastole* is a period when the auricles or ventricles respectively are not contracting.)

A How much longer is ventricular systole than auricular systole?

B i How long is ventricular diastole?

ii If this were the rate of heartbeat throughout the 24 hours, for how many of the 24 hours would the ventricular muscle be relaxing?

C After auricular systole, what causes the contracted auricular muscles to become stretched again?

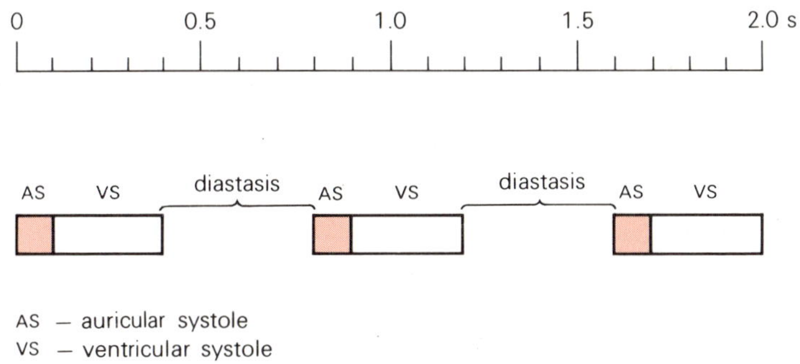

FIGURE 2·4

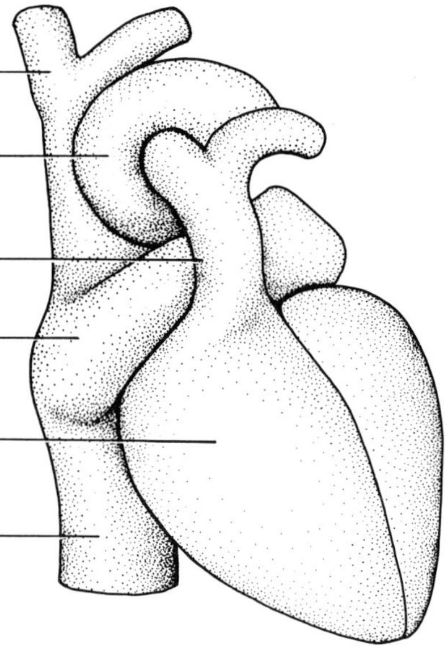

FIGURE 2·5a

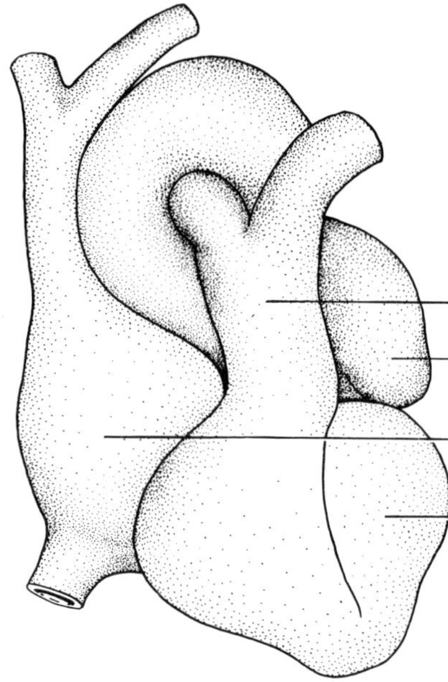

FIGURE 2·5b

D **After ventricular systole, what causes the stretched ventricular muscles to become stretched again?**

E FIGURE 2.5 **shows two views of the appearance of the human heart. One shows auricular systole, the other ventricular systole.**

State which you think is which and describe how the size of the auricles, ventricles and aorta in each drawing supports what you say.

F **The output of blood from the heart varies with the activity of the body: when more energy is needed, the heart works harder, distributing more blood and with it more sugar and oxygen for respiring cells. The average output for an adult man at rest is 5·3 l/min, but this can rise to as much as 30 l/min in a healthy young man doing vigorous exercise.**

i **The rise can be caused by increasing the number of beats per minute.**

Assuming that auricular and ventricular systole always take the same length of time, how might this increase be achieved (consult FIGURE 2.4)**?**

ii **What change in the activity of the heart could bring about increase in the volume of blood expelled at each stroke?**

G FIGURE 2.6 **shows the rate of blood flow in cm^3/min^{-1} (cm^3/min) in various parts of the human body when it is at rest and when it is in various states of activity.**

i **Copy the table and fill in the two empty columns, showing the proportions of the output of the heart which flow to the regions shown when the body is at rest** (COLUMN 5) **and when it is at maximum exertion** (COLUMN 6)**.**

ii **Taking each region in turn, describe the changes in the absolute amount of blood they receive when the body is at rest, when it is performing light exercise, when performing fairly strenuous exercise, and when at maximum exertion.**

How are the changes correlated with the changing activity of the whole body?

iii **Describe the changes in the proportional distribution to the regions when the body passes from rest to maximum exertion** (COLUMNS 5 AND 6)**.**

How are these changes correlated with the changed activity of the whole body?

REGION	1 AT REST	2 LIGHT EXERCISE	3 FAIRLY STRENUOUS EXERCISE	4 MAXIMUM EXERTION	5	6
Heart muscles	250	350	750	1000		
Skeletal muscles	1200	4500	12500	22 000		
Kidneys	1100	900	600	250		
Abdominal viscera (mainly gut)	1400	1100	600	300		
Skin	500	1500	1900	600		
Brain	750	750	750	750		
Total	5800	9500	17500	25000		

FIGURE 2·6

32

FIGURE 2.7 shows the maximum time that various vertebrates can spend under water without taking a breath. None of the animals is a fish.

ANIMAL	TIME (min)
Alligator, *Alligator mississippiensis*	120
Pigeon†	1
Hen†	3
Penguin, *Pygoscelis papua*	7
Duck (domestic)	15
White rat†	2
Rabbit†	3
Cat†	3
Dog†	4
Beaver, *Castor canadensis*†	15
Grey Seal, *Halichoerus grypus*	18
Fin whale, *Balaenoptera plupalus*	30
Bottlenose whale, *Hyperoodon rostratus*	120
Man (pearldiver)	1½

† Average time to last movement by forced submersion.

FIGURE 2·7

A **Which of the animals in the list appear to be adapted to spending part of their lives under water?**

B **We are going to try to discover how it is that these animals can remain under water for so long without taking in air. One possibility is that they might have specialized lungs or other air reservoirs, but it has been shown by dissection that their anatomy is essentially the same as that of other vertebrates. The needs of their cells are also the same, in particular the brain and heart must receive constant supplies of oxygen. If the brain is starved of oxygen for only a few minutes it is fatal to the animal.**

ANIMAL	1 TOTAL OXYGEN AVAILABLE IN THE BODY	2 O_2-CONSUMPTION WHEN RESTING	3 LONGEST PERIOD FOR WHICH ANIMAL COULD DIVE IF O_2 USED AT SAME RATE AS COLUMN 2 (min)
Alligator, *Alligator mississippiensis*	60 cm^3	4 cm^3/min	*c.* 15
Duck (domestic)	85 cm^3	22 cm^3/min	*c.* 4
Penguin, *Pygoscelis papua*	270 cm^3	100 cm^3/min	*c.* 3
Fin whale, *Balaenoptera plupalis*	3350 l	200 l/min	*c.* 17
Bottlenose whale, *Hyperoodon rostratus*	109 l	3–4 l/min	*c.* 36

FIGURE 2·8

Here are four facts about diving vertebrates:

I **Many diving vertebrates, especially very deep divers (seals, whales), far from having extra large lungs in which to store air, expire large quantities of air just before they dive or while they are diving.**

i **How might this assist the animal in diving to depth?**

II **The percentage of body weight due to blood in natural divers is in most cases about twice that of non-divers.**

ii **How might this be of benefit to the animal when diving?**

III **The blood of non-divers may carry about one quarter of its own volume of oxygen. For the natural divers, the figure may be as high as one half.**

iii **How may this be of benefit to the animal when diving?**

IV **Diving vertebrates have very dark muscles, sometimes almost black. This is because they contain a great deal of myoglobin.**

iv **How may this be of benefit to the animal when diving?**

C **Examine** FIGURE 2.8. **It shows (**COLUMN 1**) the total amount of oxygen known to be available during diving in the bodies of five of the vertebrates listed in** FIGURE 2.7.

COLUMN 2 shows the amount of oxygen used per minute by the animal when it is at rest and breathing air.

In COLUMN 3 the figure in COLUMN 1 has been divided by that in COLUMN 2, giving the maximum length of time for which we might expect the animal to remain submerged.

Compare the figures in COLUMN 3 with the corresponding figures for the same animals in FIGURE 2.7.

i **What is striking about the comparison?**

ii **Assuming the rate of oxygen consumption is the same during diving as it is when the animal is resting (it is probably more), what comment does this comparison throw on the answers to questions B ii, iii and iv ?**

33

We see then that the oxygen reserves of these animals are insufficient to explain the observed duration of their dives, assuming that all the tissues of the animal's body continue to use oxygen at the same rate during diving as when resting.

A **A clue to how the animals make the oxygen they have last out for the duration of the dive is given by looking back to FIGURE 2.1.**

The essential thing for the animal to do during diving is to maintain the oxygen supply to the brain and the heart. If blood carrying oxyhaemoglobin is passing through capillaries close to muscle cells containing myoglobin, will the myoglobin give oxygen to the haemoglobin or will the haemoglobin given oxygen to the myoglobin?

Explain how FIGURE 2.1 gives you your answer.

B **FIGURE 2.9 shows**

i **the decrease in oxygen content of myoglobin of the muscle and**

ii **the decrease in oxygen content of the arterial blood of a seal during diving.**

Compare the proportions of oxygen remaining in the blood and in the muscles after ten minutes' diving. How does the comparison show that the arterial blood does not give oxygen to the muscles?

C **Remember that myoglobin has a higher affinity for oxygen than does haemoglobin. Yet during diving, the blood does not give oxygen to the muscles, which are using it to produce energy in respiration.**

What possibility does this indicate concerning the circulation of the blood through the muscles when diving?

34

When glucose is oxidized in cells it forms carbon dioxide and water as waste products. Energy is liberated, a large part of which is bound up in ATP for use in chemical reactions in the cell. This process can be summarized as:

1 molecule glucose + 6 molecules oxygen → $6CO_2 + 6H_2O$ + energy.

If there is insufficient oxygen available, glucose can be converted into lactic acid instead. Some energy is liberated, though not as much as in the previous reaction (see question 22 **D** p 27). This process can be summarized as:

1 molecule glucose → 2 molecules lactic acid + energy.

Thus the formation of lactic acid in cells, and so its appearance in blood leaving the cells, is a measure of the oxygen starvation of the cells.

FIGURE 2.10 shows the changes in lactic acid concentration of the blood of a seal during diving (the same dive that gave FIGURE 2.9). At the start of the dive there is a little lactic acid present in both muscle and blood. Assume that the animal is equally active in each part of the twenty minute diving period.

A **How is the change in lactic acid concentration of the muscle by the end of the first five minutes of diving related to the amount of oxygen available in the oxymyoglobin in the muscle (line 2 of FIGURE 2.9)?**

B **How is the change in lactic acid concentration of the muscle by the end of**

i **ten minutes**

ii **fifteen minutes of diving related to the amount of oxygen available in the muscle?**

C **What has happened to the lactic acid content of the blood during this fifteen minutes? How is this related to the amount of oxygen available in the blood during this time (line 1 of FIGURE 2.9)?**

D **How does FIGURE 2.10 indicate that it is unlikely that blood is circulating through the muscle during the dive?**

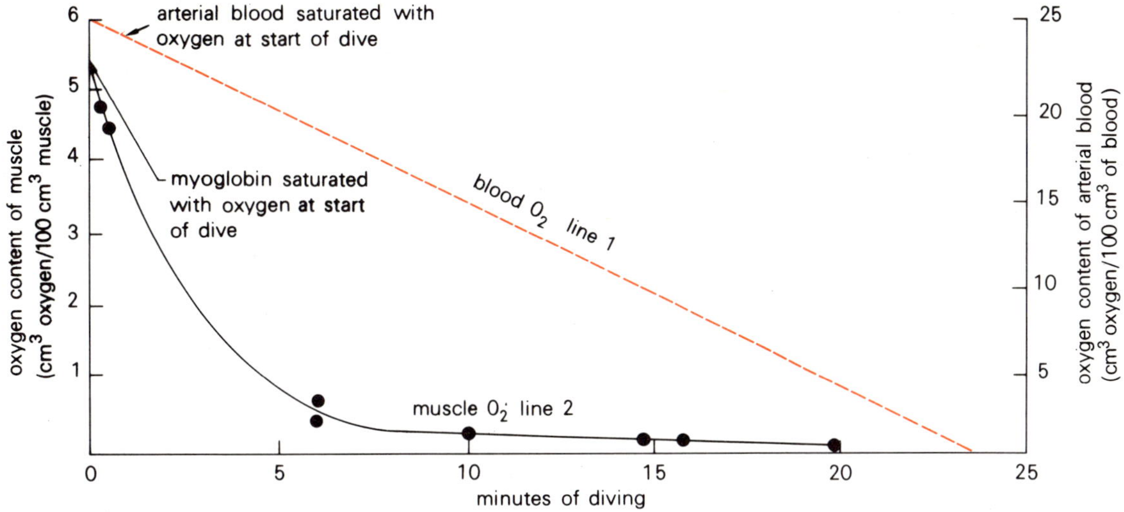

FIGURE 2·9

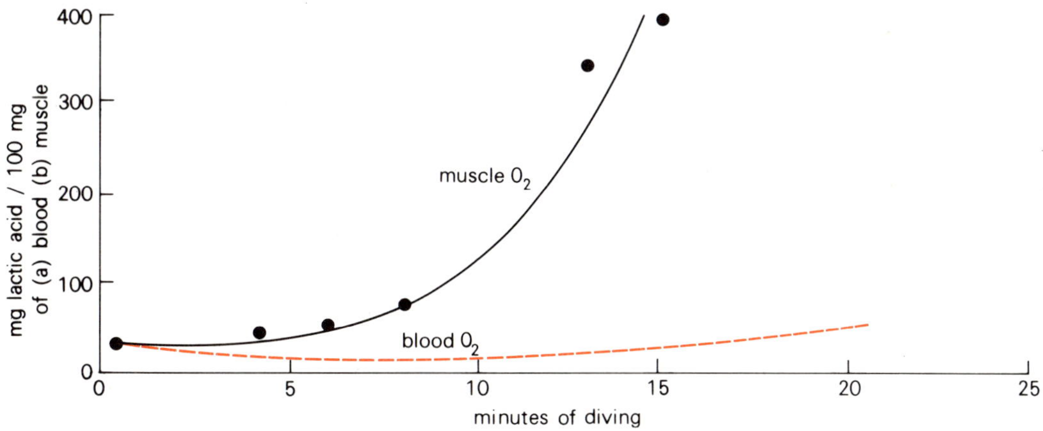

FIGURE 2·10

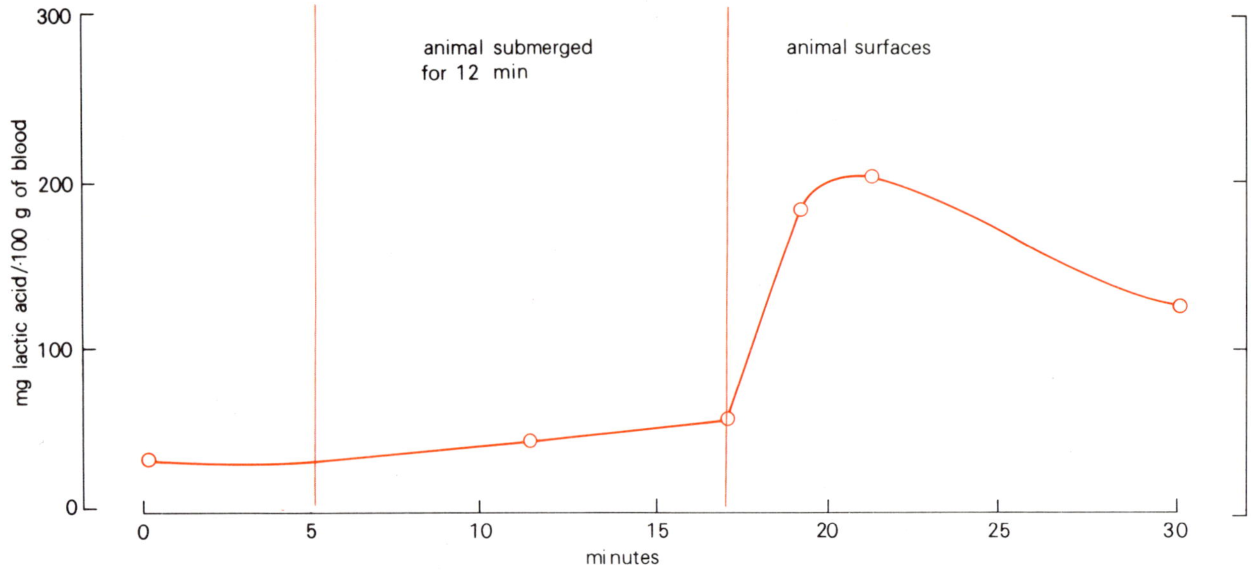

FIGURE 2·11

35

When the animal surfaces from its dive, there is a rapid change in the lactic acid content of the blood. This is shown in FIGURE 2.11, which also gives the results of an experiment with a seal, though the dive was a different one from that which gave us FIGURES 2.9 and 2.10.

A **By referring to the concentrations of lactic acid in the blood**
- i **at the start of the dive,**
- ii **after fifteen minutes' diving,**
- iii **after twenty minutes (three minutes after surfacing),**
- iv **after thirty minutes, describe the pattern of change in concentration of lactic acid in the blood before, during and after the twelve minute dive.**

B **How does FIGURE 2.10 help you to explain the change in lactic acid concentration during the first three minutes after surfacing?**

C **What change do you suppose, therefore, takes place in the circulation of the seal when it surfaces after diving?**

D **If the seal is wounded in the muscle the wound will bleed profusely if the animal is breathing air. When it dives, bleeding ceases, but resumes when the animal surfaces.**

How do these observations fit in with the answers you have given to the questions on FIGURES 2.9, 2.10 and 2.11 ?

E **What mechanism do you know of that will enable the blood supply to the muscles to be shut off during a dive and opened up again after it?**

36

FIGURE 2.12 shows the effect of diving on the rate of heartbeat in the seal. Each small group of bumps on the lines was made by one heartbeat.

The figure does not show continuous results. Each line took ten seconds to make; the second line was recorded two minutes after the start of the experiment, the third line four minutes after, the

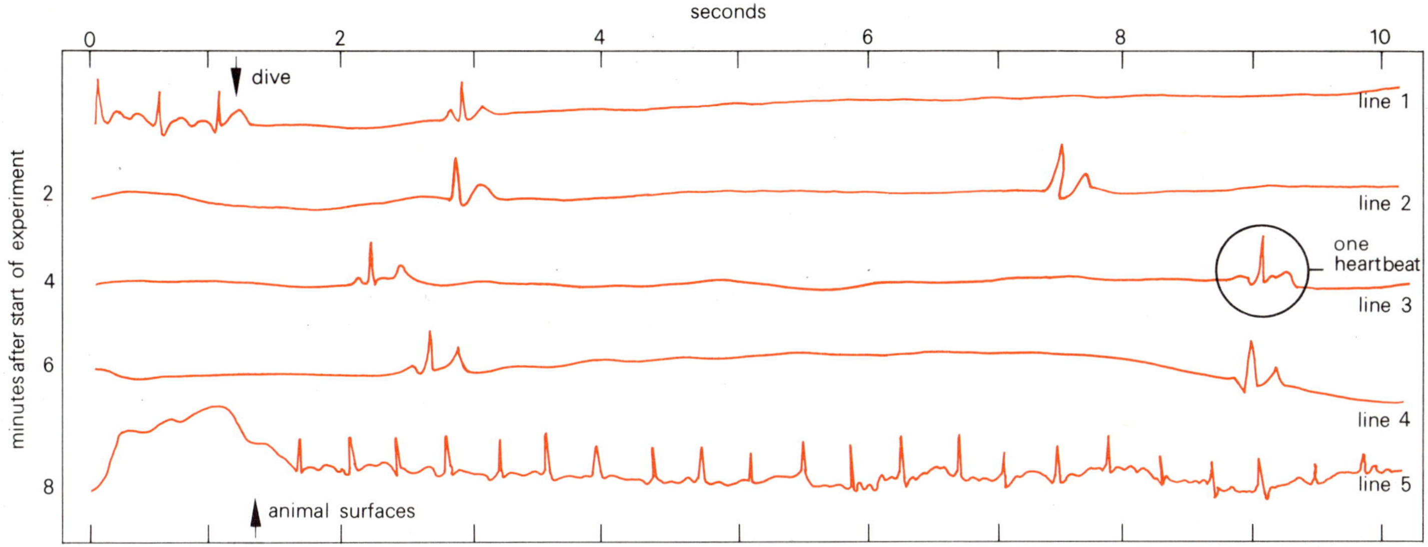

FIGURE 2·12

fifth line eight minutes after. The animal dived one second after the experiment started and surfaced eight minutes and one second after it started.

A **What was the rate of heartbeat (in beats per second)**

- **i** **before diving**
- **ii** **after surfacing?**

B **Taking lines** 2, 3 **and** 4, **how long after the first heartbeat shown on these lines is it before the next heartbeat?**

C **Assume the rate of heartbeat to be similar in the periods not shown in** FIGURE 2.12 **(e.g. the period between the end of line** 1 **and the start of line** 2**). We have discovered that during diving, the circulation to the muscles is cut off.**

How is this related to the change in rate of heartbeat that you have just described?

3 Temperature control

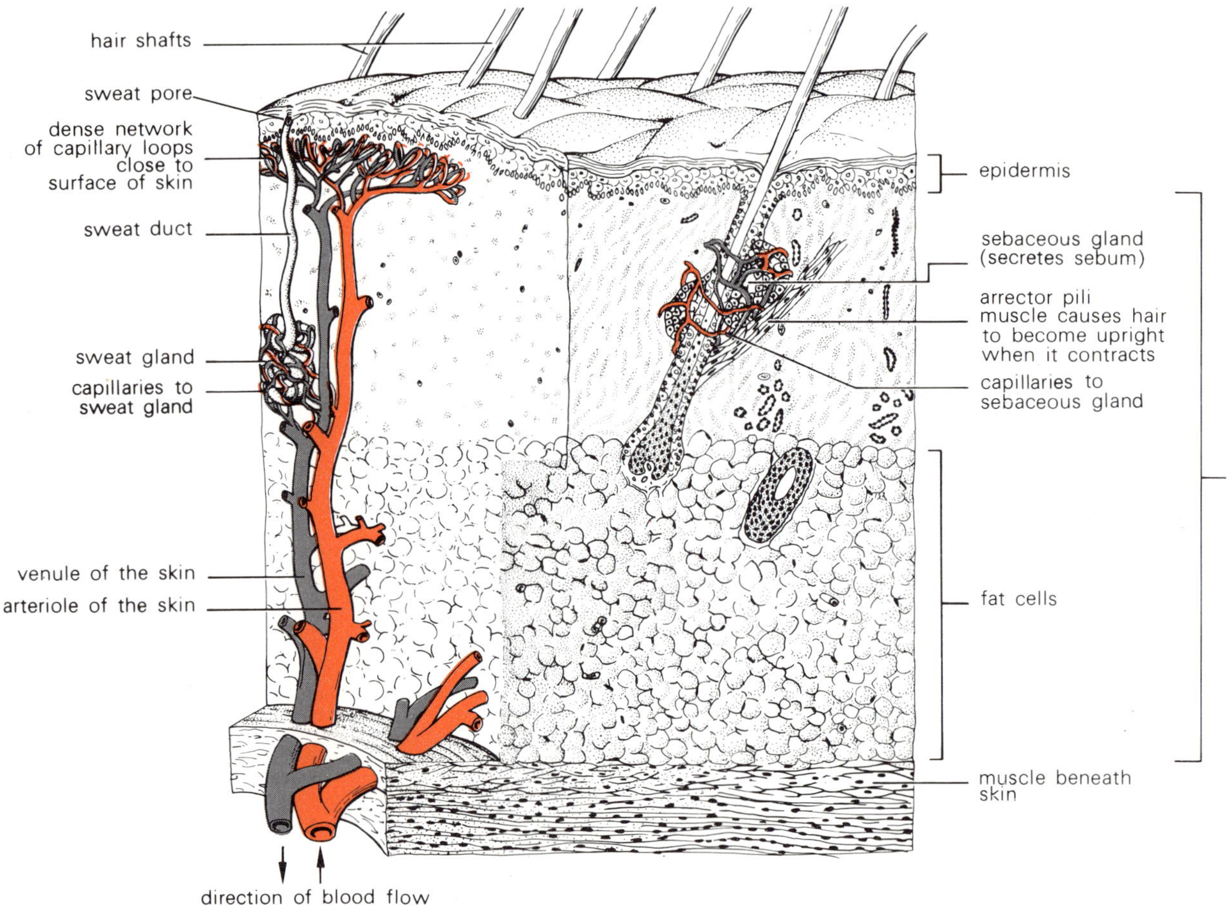

FIGURE 3·1 a

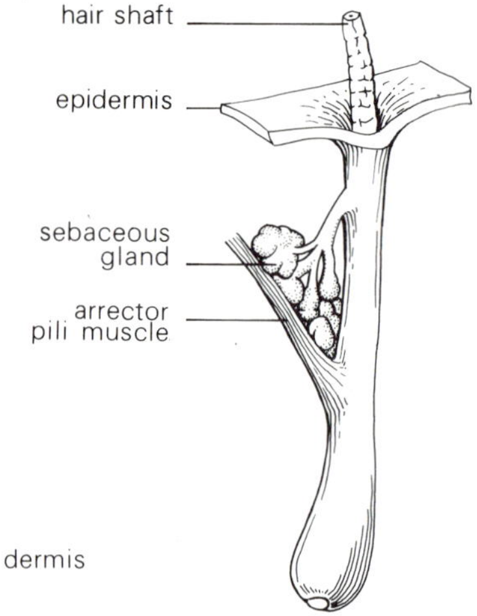

FIGURE 3.1 b

37

FIGURE 3.1 is a stereogram of a portion of human skin. The skin of other mammals is similar in structure, though there are differences in detail. For example, most mammals have a thick layer of hairs (fur), but in man only a few parts of the body bear long and dense hairs.

The Alaskan muskrat (*Ondatra zibethica*) (FIGURE 3.3), like many other aquatic mammals (e.g. beaver, mink) traps a layer of air in its hairs when swimming, and this accounts for 21% of the volume of the entire animal.

A Explain why the presence of the layer of trapped air helps to cut down heat loss from the animal's body when it is swimming in (say) ice-cold water.

B When the animal climbs out of the water it gives itself one or two shakes and fluffs out its fur. This causes nearly all the water to fly off its fur.

i How does contraction of the hair muscles cause the hairs to fluff out?

ii It is the activity of the sebaceous glands (FIGURE 3.1b) which enables the animal to rid itself of water so easily. Explain how this is so.

iii Notice the position of the sebaceous glands relative to the muscle which raises the hair. How might this position contribute to the efficient working of the glands?

iv Whales spend their whole lives in water; they have no hair. Seals are also aquatic mammals but they do emerge and bask on the land; they have fur.

How do you think these ways of life might be related to the fact that whales have no hair (except a few whiskers) while seals do?

38

Look again at FIGURE 3.1 and find the sweat gland. It consists of a tube coiled into a ball about 0·4 mm in diameter. The tube is lined with cells which produce sweat; this pours up the sweat duct on to the surface of the skin. There are about fifty sweat glands per cm² on the human palms and about 2·5 million on the whole human body surface. A man employed in a sitting-down job in a temperate climate loses about 500 cm³ of sweat per day.

A Explain the significance of the large number of capillaries surrounding the sweat gland.

B Explain why the evaporation of sweat causes the skin surface to become cooler.

C If a liquid is heated its temperature rises and it evaporates. Some of the heat is used in raising the temperature of the liquid, some in evaporating it.

If we want to know how much heat is used only in causing evaporation, we must investigate the liquid when it is boiling. Then all the heat used is causing evaporation. The temperature of the liquid does not rise above its boiling point however hard it is heated.

FIGURE 3.2 **shows how much heat is needed to turn 1 g of various boiling liquids into vapour at the same temperature.**

From the right hand column it is clear that it takes more heat to evaporate 1 g of water than to evaporate 1 g of the other liquids. Explain why this makes the evaporation of sweat an efficient method getting rid of surplus heat and at the same time of conserving body water.

D Cooling of the skin surface by sweat will cause a cooling of the whole body. What structures in FIGURE 3.1a**, apart from the glands themselves, are involved in this, and how is the general cooling brought about?**

LIQUID	NUMBER OF CALORIES NEEDED TO TURN 1 GRAMME OF THE BOILING LIQUID INTO VAPOUR AT THE SAME TEMPERATURE
Acetic acid	96·8
Aniline	110·0
Benzene	94·1
Carbon disulphide	84
Chloroform	59
Cyclohexane	85·4
Ether	83·9
Naphthalene	75·5
Nitrobenzene	79·1
Phenol	114
Pyridine	104
Water	540

FIGURE 3·2

E **What are the three ways in which the skin surface loses heat when no sweating is taking place?**

F **Explain why it is impossible or very difficult for man to live in climates which are both very hot and very humid (i.e. where the air contains a great deal of water vapour).**

39

Consider desert warfare, such as in the Arab/Israel war in the Sinai desert in June 1967. A soldier marching in the desert will absorb heat radiation from the sun. If he is wearing shorts and a shirt, those parts of him which are exposed will absorb about 240 kcal per hour. If he is marching at 5 km/hr. his body will be producing about 270 kcal per hour. He thus has a total amount of heat to lose amounting to about 510 kcal per hour.

A **Evaporation of one gramme of water as sweat requires 0·54 kcal. How much sweat must the man lose per hour in order to get rid of the heat he has absorbed and is producing?**

B **Which parts of the man's body are producing the 270 kcal per hour?**

40

Fat is a poor conductor of heat. Look at the distribution of fat cells in FIGURE 3.1a.

A **In what way will the fat affect the loss of heat from the underlying muscles?**

B **How will the fat affect loss of heat from the blood vessels in the upper and lower parts of the dermis?**

C **Whales and seals living in ice-cold water endure the greatest surface cooling of any warm blooded animal. Water and air are both poor conductors of heat, but water conducts heat about three times more efficiently than does air. Leaving aside the heat-insulating properties of fur, which whales do not have, why is the surface loss of heat from whales greater than that from, say, polar bears?**

FIGURE 3·3

D **How is the answer to C correlated with the fact that whales and seals have a very thick layer of fat (blubber) in their dermises?**

E **An active large whale (which may be up to 36 m long) must produce a great deal of heat. Its surface area relative to its bulk is small. If whale and seal skin have the same basic pattern as human skin (FIGURE 3.1), will the layer of blubber prevent the whale from regulating its body temperature via its skin? Explain your answer.**

41

We saw in question 39 that a man marching in the desert heat may lose up to a litre of sweat per hour compared with the 500 cm^3 lost per day by a man in a sitting-down job in a temperate climate. Two things raise the man's body temperature, radiation absorbed by his skin and heat produced in his muscles.

Sweat glands are supplied with nerves which cause them to secrete more or less sweat, or none.

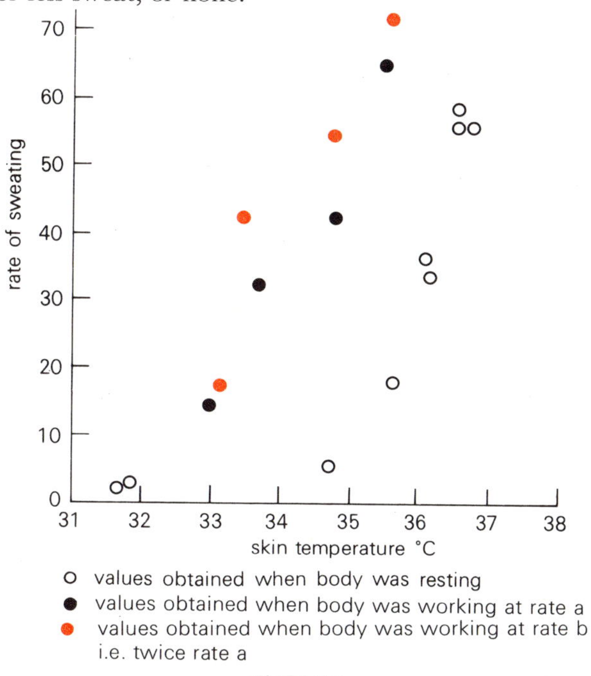

FIGURE 3·4

Mammals respond to changes in their environment by means of a fairly standard pattern of events. Sensitive cells detect changes in the environment, pass this information on to nerves and perhaps to the spinal cord or the brain; this sends messages to other cells or organs causing a change in behaviour to meet the change in the environment. Thus if one picks up a hot dish, temperature-sensitive structures in the skin detect the change in temperature. They pass the information to nerve cells which take it to the spinal cord. Messages from here may cause the arm muscles to contract, resulting in the opening of the fingers and the dropping of the plate. This simple reflex may be overruled by the brain, and one can retain hold of the plate if one wishes to.

In a similar way, there is presumably a temperature-detector (a thermostat) in the body which detects rise in the body temperature and passes the information to nerve cells which cause the sweat glands to produce sweat and so cool the body. Since a rise in body temperature can be caused either by absorbing heat from outside the body or by producing heat inside it, this 'thermostat' might be situated in the skin or somewhere inside the body.

FIGURE 3.4 is a graph showing how the rate of sweating is related to skin temperature under three sets of experimental conditions

a when the body was resting

b when the body was doing a certain amount of work

c when the body was doing twice as much work as in b.

The subject of these experiments (a man) lay naked, suspended and totally enclosed in a specially constructed chamber which was able to measure very accurately the amount of heat being given off by his body.

A **Look at the rates of sweating under the following conditions:**

i **body resting, skin temperature 34·8°C**

ii **body working at rate** a, **skin temperature 33·0°C**

iii **body working at rate** b, **skin temperature 33·5°C**

What do these readings, and other similar ones you may make from FIGURE 3.4, tell you about the relationship between skin temperature for all three experiments taken together and rate of sweating?

B **What does this suggest about the possibility of the temperature detector controlling sweating being located in the skin?**

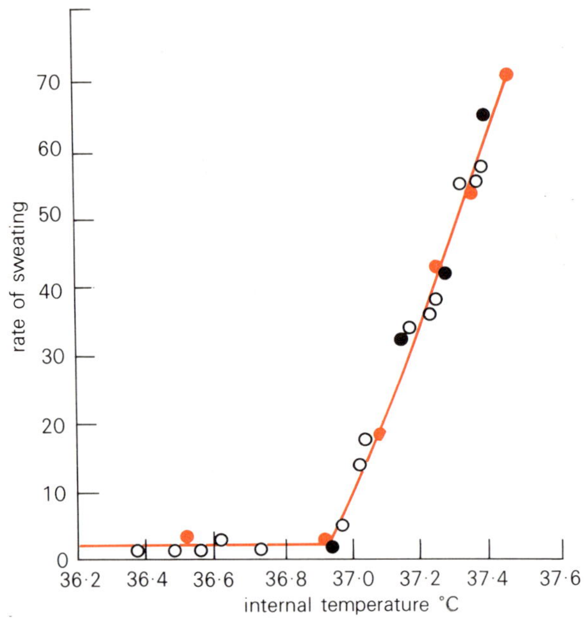

FIGURE 3·5

42

FIGURE 3.5 is a graph showing how the rate of sweating is related to internal body temperature. The site chosen was the ear drum, which is close to the middle of the underside of the brain.

The range of internal temperature that can be brought about (e.g. by eating ice or by moving the muscles) is narrower than the range of skin temperatures in FIGURE 3.4, but this time the points are not scattered, they form a clear pattern.

A **Are the points recorded on this graph when the body was resting or working at rate a or rate b related to each other? Compare FIGURE 3.5 with FIGURE 3.4 and explain how the comparison gives you your answer.**

B **Describe the shape of the graph and say what it tells you about the relationship between internal body temperature and rate of sweating.**

C **What does this graph suggest about the probability of a temperature-detector controlling sweating being situated near to the eardrum?**

43

FIGURE 3.6 shows more results from the same series of experiments. The man rested in his temperature-controlled chamber which was kept at 45°C. At the times shown he ate a quantity of crushed ice. His internal body temperature was taken at his eardrum.

A **What was the man's internal body temperature over the first 25 minutes of the experiment (graph A)? What was his rate of sweating over this period (graph B)? Was he gaining or losing heat over this period?**

B **What happened to the following over the next 25 minutes, after the man had eaten his crushed ice?**

- **i** **the internal body temperature**
- **ii** **the skin temperature**
- **iii** **the rate of sweating**
- **iv** **the rate of heat loss from his body**

C **During the first 10 minutes after eating the ice, the skin temperature rose sharply.**

- **i** **Why was this?**
- **ii** **Did this cause an increase in sweating?**
- **iii** **What does this tell you about the probability of the temperature detector (the thermostat) controlling sweating being situated in the skin?**

D **How is the change in internal body temperature, as measured at the ear drum, during the 25 minute period after eating the ice related to the rate of sweating? What does this indicate about the situation of the temperature-detector controlling sweating?**

E **Explain how the change in rate of heat loss from the man's body over the 25 minute period after eating ice is related to the change in rate of sweating in the same period.**

F **The results of these experiments indicate that the temperature detector controlling sweating is situated near to the ear drum. In fact it is in a small part of the under side of**

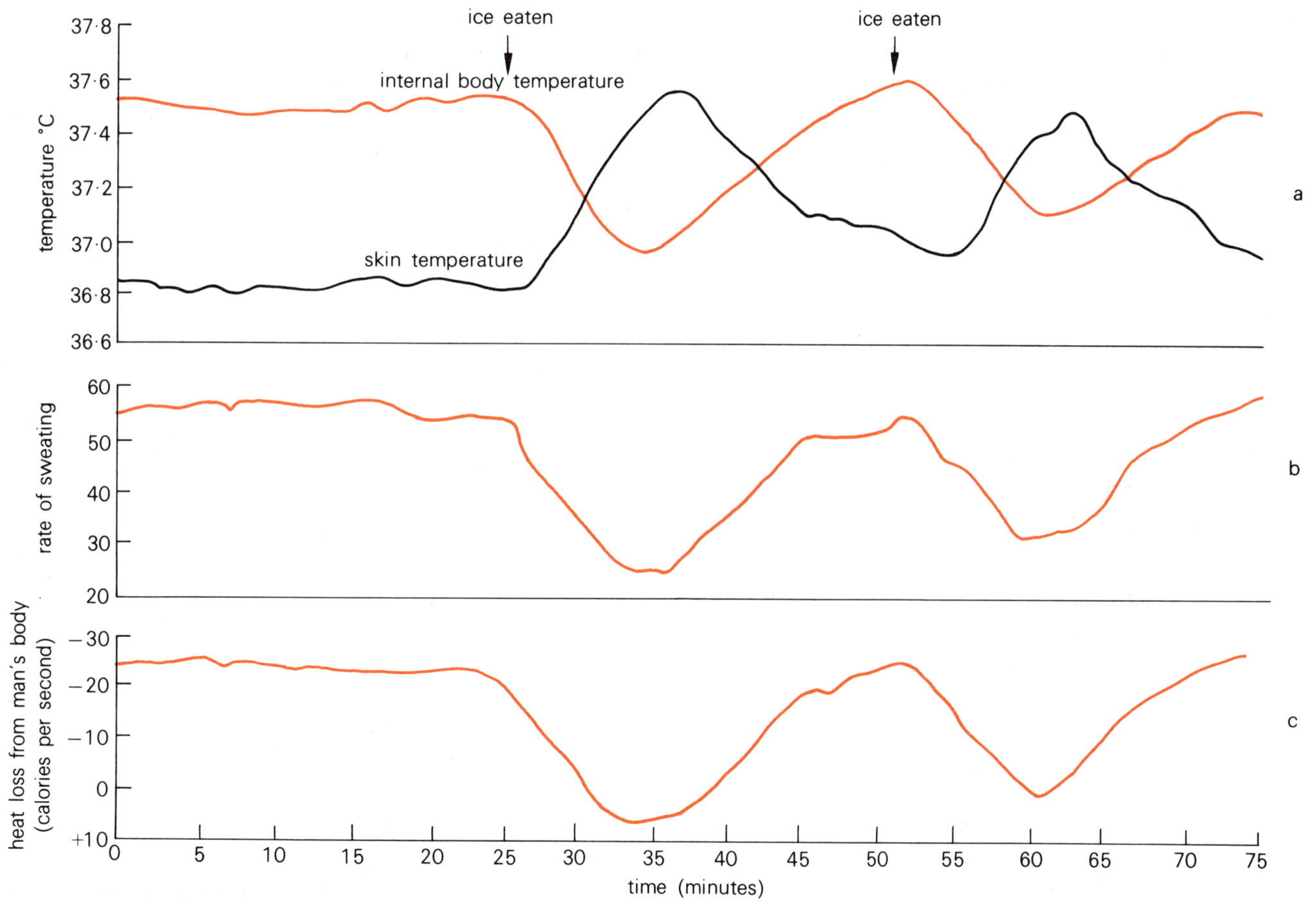

ice eaten
ice eaten
internal body temperature
skin temperature
temperature °C
37·8
37·6
37·4
37·2
37·0
36·8
36·6
a
rate of sweating
60
50
40
30
20
b
heat loss from man's body (calories per second)
−30
−20
−10
0
+10
c
0
5
10
15
20
25
30
35
40
45
50
55
60
65
70
75
time (minutes)

FIGURE 3·6

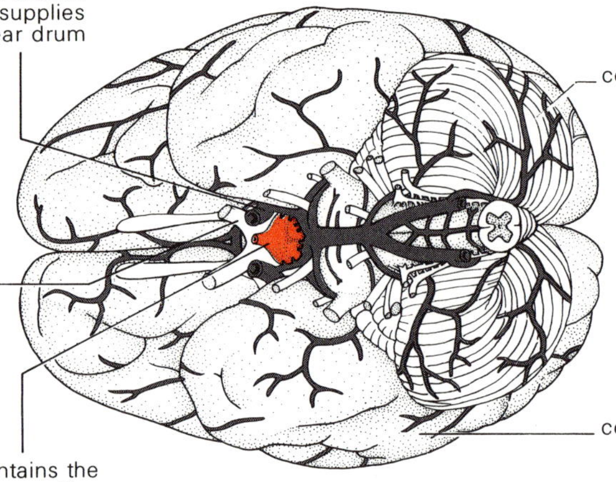

FIGURE 3·7

the brain, called the hypothalamus. **You can see the hypothalamus in** FIGURE 3.7.

The internal body temperature in this experiment was measured at the ear drum, but when the man ate the ice it was his stomach which was cooled first. How did the change of temperature in his stomach become passed on to the hypothalamus, causing it to set off the nerve impulses making the sweat glands produce less sweat?

What feature of FIGURE 3.7 **lends weight to your answer?**

SIDE cm	SURFACE AREA cm^2	VOLUME cm^3	SURFACE AREA/VOLUME
1	6	1	6:1
2	X	X	X:1
3	X	X	X:1
4	X	X	X:1
5	X	X	X:1
10	X	X	X:1
25	X	X	X:1

FIGURE 3·8

44

FIGURE 3.8, when completed, shows the relationship between the surface area and volume of a series of cubes of side 1, 2, 3, 4, 5, 10 and 25 cm.

A Copy the table and fill in the blanks (each has been marked with a cross).

B How does the ratio of surface area to volume change as the cubes increase in size?

C Consider instead of cubes two mammals, a whale and a harvest mouse. Which has the larger surface area relative to its volume?

45

The African elephant lives in a hot climate.

I It has practically no hair.

II It is the largest living land animal and so has a small surface area relative to its bulk.

III When it raises its ears, its total surface area is increased by about one sixth.

How is each of these facts related to the elephant's ability to control its body temperature?

46

FIGURE 3.9a is a drawing of a fruit bat (*Epomops buttikoferi*) hanging from a perch. FIGURE 3.9b is a drawing of the same animal with its wings folded. The animal's body is about 16 cm long from the head to the top of the leg; the wing span is about 45 cm.

A Disregard the wings for the moment. This is a small mammal. What can be said about the surface area of its body relative to its bulk, and how will this affect the rate at which it loses heat to its surroundings?

B Work out how much the surface area of the animal increases when it holds its wings out fully. FIGURE 3.9a **has been superimposed on a grid to enable you to make a rough calculation of the area of the wings (remember that each wing has two surfaces, upper and lower). To find the approximate area of the body, suppose it to be a cylinder of**

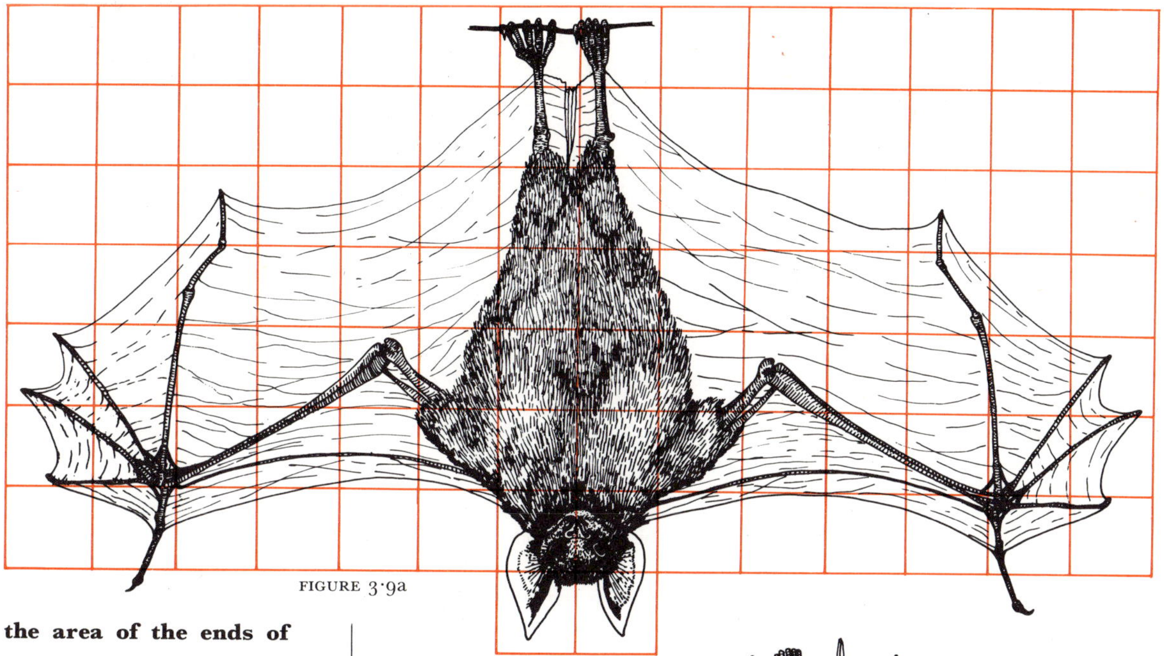

FIGURE 3·9a

circumference 10 cm, and ignore the area of the ends of the cylinder.

C **What will be the effect on heat loss from the animal of opening and flapping the wings?**

D **What do** FIGURES 3.9a **and** b **reveal about the means which the animal has of conserving its body heat?**

E **This small mammal has no control over the amount of heat which its body produces. When flying, its body temperature is high, but when it rests it falls rapidly and after an hour its body is at the temperature of its surroundings.**

Explain why, on waking from rest, the animal cannot fly off but can only make small movements like squeaking, biting and crawling. In answering, consider the effect of temperature on the activity of enzymes, and so on the activity of cells generally.

F **In order to raise its body temperature sufficiently to be able to fly (to about 30°C), the bat, staying in the position shown in** FIGURE 3.9b, **undergoes a series of physical jerks, shivering and raising its body by flexing its legs. Give two reasons why it is important for the bat to remain wrapped up in its wings while doing this.**

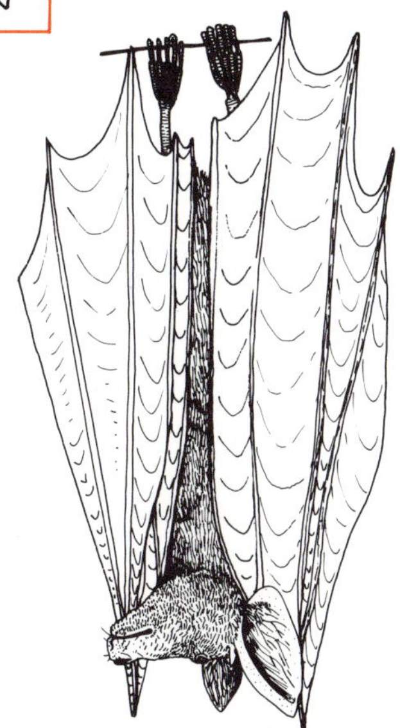

FIGURE 3·9b

47

Look back to FIGURE 3.3, the Alaskan muskrat (p 42). You may know its thick brown fur, which is used for making musquash fur coats. The animal's body is 25-35 cm long. Its flattened tail, which it uses for swimming, is 20-28 cm long, and is either very thinly furred or naked. It is a water rat and is distributed over most of North America. It spends a good deal of the year swimming in ice-cold water.

We have already noted that the thick layer of body fur traps air when the animal is swimming, the volume of which is equal to about 21% of the volume of the entire animal. FIGURE 3.10 enables us to see how effective this is in cutting down loss of body heat.

Line A shows the temperature of the animal's rectum and line B the temperature beneath its skin in the trunk region when it passes from air of temperature 18-20°C to ice water (2°C) and back into air.

A **How long did the animal spend in the ice water in this experiment?**

B **What was the effect of passing from air to ice water and back to air on the rectal temperature and on the temperature beneath the skin in the trunk region?**

C **Apart from its fur, what other heat-insulating mechanisms would you expect this animal to have?**

D **The surface area of the tail forms a considerable part of the total surface area of the animal's body. It has very little or no hair on it. What would you suppose would be the consequence of these two facts on heat loss when the animal is swimming in ice-cold water?**

E **Line C in FIGURE 3.10 shows us what happened to the temperature of the tail when it was immersed in ice water after having been kept in air at 18–20°C. (The rest of the animal was held still during this part of the experiment.)**

i **How did the temperature of the tail** in air **compare with that of the rectum and the trunk region under the skin?**

ii **Describe what happened to the tail temperature when the tail was immersed in ice water and for the first 40 minutes after removal from it.**

iii **The animal needs its large tail in order to swim. How is the behaviour of the tail temperature you have just observed of great benefit to the animal? What would you say about the property of the tail tissues (e.g. muscle, nerve, skin) compared with those of the rest of the animal's body?**

iv **We saw in question B that the animal's fur coat provides a good heat insulator. Maintaining a constant body temperature is brought about here as in other mammals by a balance between heat production and conservation in the body and heat loss from it.**

If the animal needed to lose heat, it could not lose much from the body surface, but it might be able to lose it from its tail, which is almost naked. Look at the point on line C in FIGURE 3.10 where the animal is stimulated into exercise (remember that it had previously been held still in this experiment).

What does the graph tell you about the possibility of the tail playing a part in heat loss as well as in heat conservation?

48

When the muskrat is swimming in ice-cold water, its tail is very active. Its muscles require energy, which they obtain from glucose and oxygen. Therefore a good blood supply to the tail must be maintained.

Thus the fact that the tail becomes ice-cold when it is cooled is not because the supply of warm blood to it is cut off. Yet the warm blood entering the tail neither prevents the tail from becoming cold nor warms it when it has become cold.

This is brought about by the arrangement of the blood vessels entering and leaving the tail. A large number of arteries enter it, and each is surrounded by a thin-walled vein leading back into the body (FIGURE 3.11).

Explain how this arrangement of arteries and veins will prevent heat from the warm arterial blood from reaching far into the tail, where it would be lost to the surface and the surrounding water.

49

There is a heat-losing arrangement similar to that of the muskrat tail in the rabbit ear. This is large, thin and well supplied with

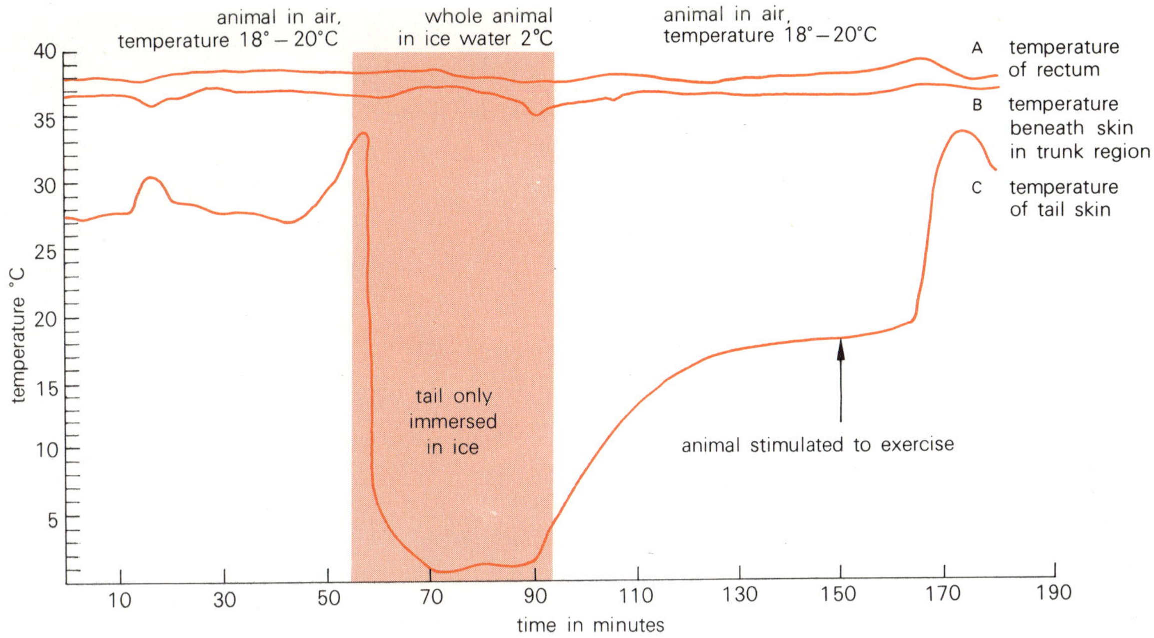

FIGURE 3·10

blood. The arrangement of blood vessels is similar to that of the muskrat tail (a 'counter-current' arrangement, where the out-flowing 'current' in the arteries comes close to the inflowing 'current' in the veins). Heat is passed from warm arterial blood entering the ear to the cold venous blood returning from it, so in cold conditions not much heat is lost from the surface of the ear.

It was shown in one experiment that every cubic centimetre of blood entering the ear gave up 22 calories to the heat exchanger (passing back to the body in the venous blood) and only 5 calories to the ear surface and the environment.

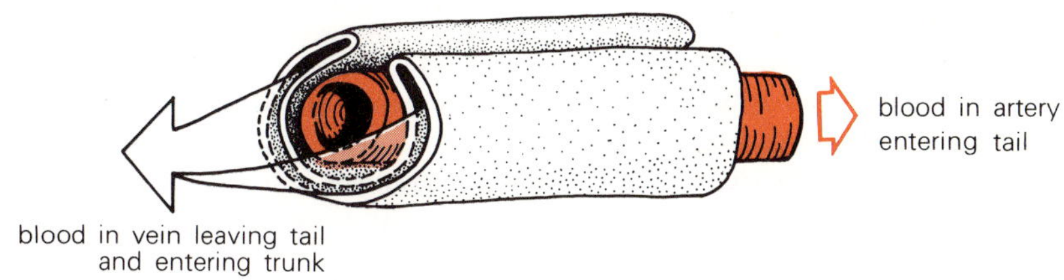

FIGURE 3·11

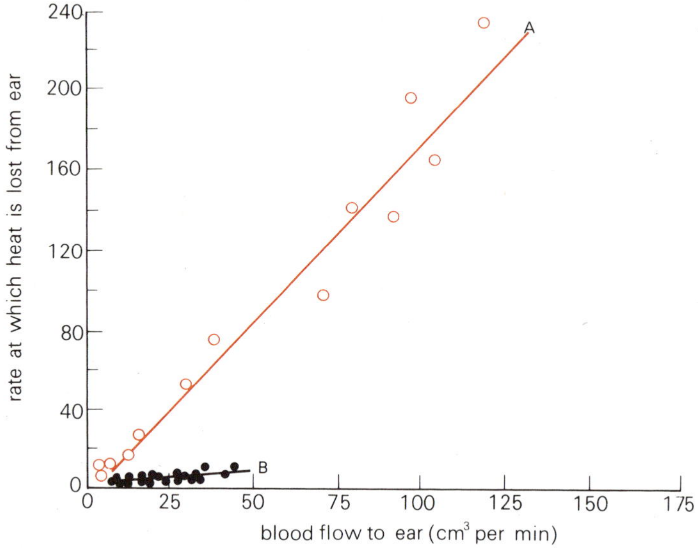

FIGURE 3·12

FIGURE 3.12 shows the results of two experiments in which rate of loss of heat from the rabbit ear (Y axis) was measured when the rate of flow of blood to the ear was varied (X axis).

In the first experiment the temperature of the air round the animal was 23°C (line A); in the second the air temperature was 4·6°C (line B). 23°C is a warm summer temperature; 4·6°C is a cold winter temperature.

A **The range of rates of flow of blood in the ear obtained when the temperature of the air round the animal was 4·6°C was much lower than the range obtained when the air temperature was 23°C. Suggest an explanation of this fact.**

B **Explain how the slope of line A (the angle made by the line with the X axis) and its length tell you that when the air temperature is 23°C the ear is acting as a cooling organ.**

C **Explain how the slope and length of line B tell you that when the air temperature is 4·6°C, the heat exchanger has been called into play and the ear has cut down its heat loss a great deal.**

FIGURE 3·13

D **Taking one particular point on the graph, how does the rate of heat loss in the two experiments when the rate of flow of blood is 50 cm^3 min^{-1}, confirm your answer to C ?**

50

Although at first sight it seems unlikely, it is a fact that many mammals are capable of living permanently in very dry and very hot deserts. There are large populations of desert rodents such as the desert rat, the gerbil, the kangaroo rat (FIGURE 3.13) and the pocket mouse; and a dependent population of carnivorous mammals feeding on them.

Most of these creatures have no special ability to resist high temperatures. They live underground and are active only at night after the sun has gone down. Thus they do not squander water in keeping themselves cool by evaporation. We will see where they obtain their water in question 57.

The camel is a large mammal which cannot escape the heat of the day. In the Sahara desert, camels (the dromedary camel, *Camelus dromidarius*) feed in the winter on sparse vegetation, which

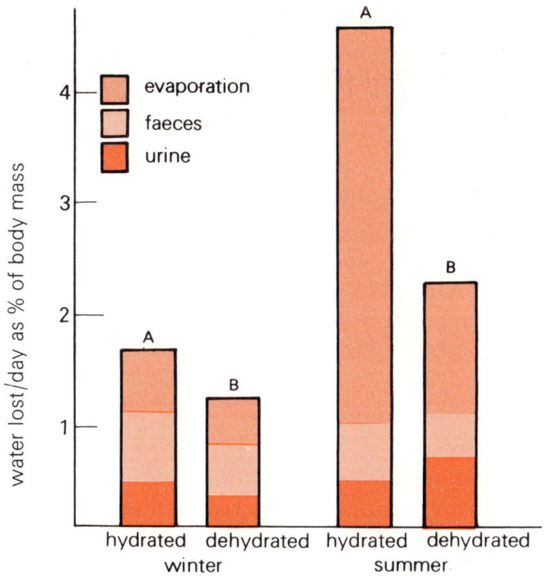

FIGURE 3·14

contains enough water to fulfil their needs without the animal drinking. In the summer, however, much of the vegetation dries up, and at the same time the camel's need for water increases for it sweats in order to prevent its body from becoming excessively hot. In the summer the Arabs supplement the very sparse natural diet with dates and hay, and the animal drinks about once in three days. A camel can go for as long as seventeen days without water, eating only dry food, and exposed to daytime temperatures of over 40°C.

We are going to look at some of the mechanisms which the camel possesses which enable it to tolerate life in such unpromising conditions. First we will examine its water loss in urine, faeces, and evaporation.

FIGURE 3.14 shows the amount of water lost per day in urine, faeces, and by evaporation when the animal was:

A given water frequently

B given no water.

Values are shown for winter, when the air temperature was about 20°C, and summer, when the air temperature was about 40°C.

A What percentage of the animal's body weight is lost per day as water in the summer when the animal is given water frequently?

B What can be said about the loss of water in both winter and summer when the animal is deprived of water compared with the loss when it is given water freely?

C Is the loss from each of the three sources, urine, faeces, and evaporation, less in the winter when the animal is deprived of water than when it is given water freely?

D In the summer the animal had a more salty diet than that in the winter. How is this reflected in the amount of urine produced in the summer?

E What is the greatest cause of increased water loss from the watered animal in the summer?

F What is the most effective factor in cutting down water loss in the summer when the animal is deprived of water?

51

If too much water is lost from the body, the blood becomes too thick (viscous), the circulation fails and the animal dies. In most mammals, including man, this happens when about 12% of the body weight has been lost as water.

A The camel which gave FIGURE 3.14 **was kept in the summer heat of the Sahara for seventeen days without water. It showed no ill effects and was quite healthy at the end of this time. What percentage of its body weight had it lost as water at the end of the seventeen days?**

B A man weighing 64 kg exposed to the same conditions would lose about 1 kg of water per hour as sweat alone. How long would it take him to lose 12% of his body weight as water?

52

The dromedary camel is able to drink very rapidly. After a period without water it can drink 20% of its own body weight of water in ten minutes. Today dromedary camels are domestic animals. There are some wild bactrian (two-humped) camels, *Camelus bactrianus*, in the Gobi desert in central Asia. Another member of the camel family, the Guanaco (*Lama guanicoe*) of South America, is entirely wild.

Why would it be an advantage for a wild camel to be able to drink water so quickly?

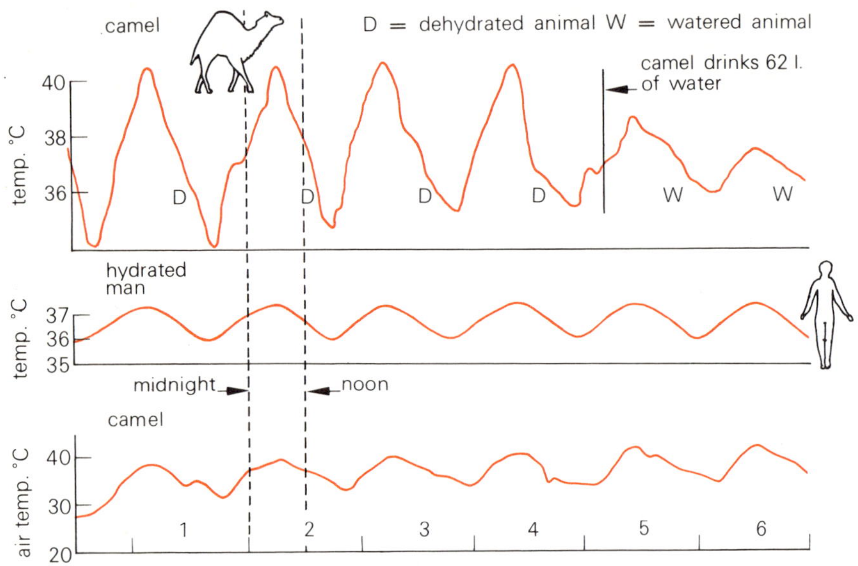

FIGURE 3·15

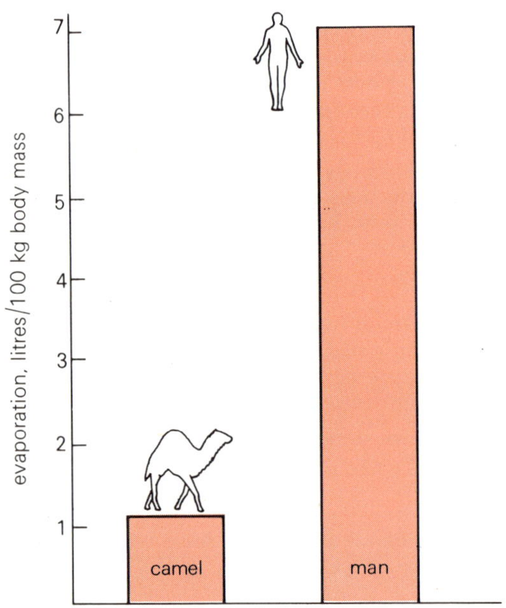

FIGURE 3·16

We have seen in questions 50, 51 and 52 that the camel loses less water when it is deprived of water, in particular it loses less by evaporation in the summer; that it can tolerate loss of water equal to 30% of its own body weight with no ill effects; and that it can rapidly replace its water loss by drinking. All these are advantages helping the camel to tolerate desert conditions.

FIGURE 3.15 reveals another special feature of the camel's physiology which has important survival value. The top line shows the fluctuation of the animal's body temperature (measured in the rectum) day by day at the end of a long period during which it had not drunk. It was resting (not working), fully exposed to the Sahara sun each day.

In the early part of the fifth day shown on this graph the animal drank 62 litres of water.

Also shown on the graph are the daily fluctuations of air temperature and the daily fluctuations of the rectal temperature of a man measured under the same conditions, though he was allowed to drink as much water as he needed.

A **What was the range of temperature fluctuation in the man's body per day?**

B **What was the range of temperature fluctuation in the environment per day?**

C **Is the range you obtained in A what you would expect in view of the environmental temperature range observed in B? Give the reason for your answer.**

D **What is the maximum range of temperature fluctuation observed in the camel when it had been deprived of water?**

E **We see that unlike the body temperature of man, the body temperature of the camel can rise as much as 6°C. The consequent difference in the amount of sweat lost from the two animals is shown in FIGURE 3.16. What is the difference, and how does it contribute to the animal's ability to cope with the desert environment?**

F **The temperature fluctuation of the camel's body is parallel to that of the environment surrounding the animal. When the environmental temperature is highest, so is that of the camel's body. When the environmental temperature is lowest, so is that of the camel's body. Check this in FIGURE 3.15.**

The camel stores the heat it receives during the day.

i How does it get rid of this heat?

ii Why is this method of getting rid of heat helpful in conserving body water compared with the method used by man?

G When the animal drinks (beginning of day 5 on FIGURE 3.15 **), there is an immediate change in the pattern of its daily body temperature fluctuation.**

i What is this change?

ii How is it related to the water loss shown in FIGURE 3.14 **when the animal is drinking regularly in the summer?**

54

FIGURE 3.17 shows that the dromedary camel has quite a thick fur coat. On the animal's flanks the fur is about 3 cm thick, in the mid line about 5-6 cm thick, and on the hump up to 11 cm thick.

In most environments, mammals are warmer than their surroundings, and the fur and the air trapped in it are used as a barrier of poor heat-conducting material helping to cut down heat loss. In the case of the camel the position is reversed. The environment, as we have seen, is hotter than the animal, and the fur and the air it contains may be expected to act as a barrier preventing heat from passing into the animal.

FIGURE 3.18 shows how effective the fur of the camel is in reducing the animal's water loss. In *experiment I* camel A had its fur on, camel B was closely shorn. In *experiment II* both animals were closely shorn.

A Compare the water loss between the two animals in *experiment I* **and between the two animals in** *experiment II* **and say what effect the fur had on water loss.**

B Camel A **weighed about 250 kg; camel** B **weighed about 500 kg. In** *experiment I*, **how much more water did camel** B **lose per day than camel** A **?**

55

Make a list of all the adaptations you have learned of which enable the camel to live in the desert environment.

FIGURE 3·17

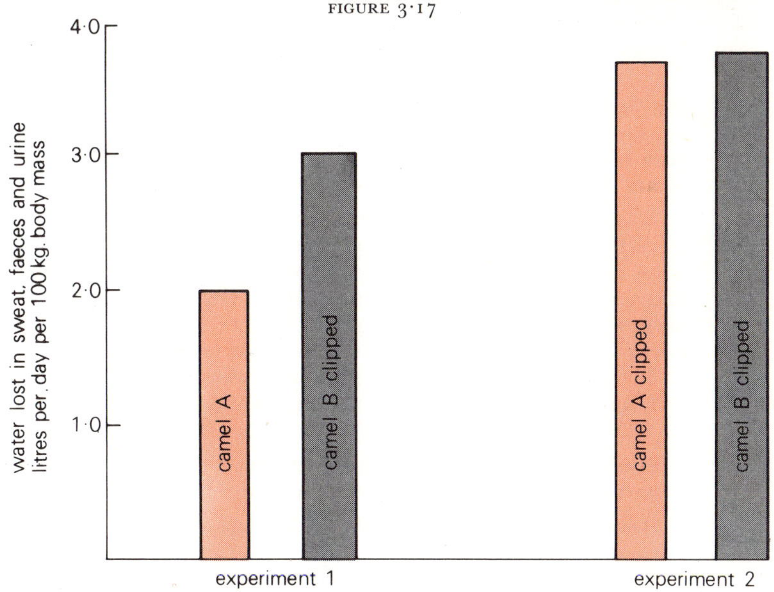

FIGURE 3·18

56

Most small mammals living in deserts escape the problems imposed by the hot environment. They live in burrows in the day and emerge only in the cool of the night.

Only one desert mammal weighing 100 g or less has been studied from the point of view of its temperature regulation, namely the antelope ground squirrel, *Citellus leucurus* (FIGURE 3.19). This animal does not follow the general rule given in the paragraph above. It is active during the day. It does not aestivate (that is, spend the summer in a dormant condition) as some desert rodents do, but is active at all seasons. Its body temperature when it has rested for some time in its burrow is 36°C. The air temperature outside the burrow may be up to 42°C. If it were to try to keep its body temperature down to 36°C by sweating it would have to evaporate an amount of water equal to 13% of its body weight per hour. Since the animal may never have access to drinking water in the whole of its life, this is obviously impossible.

FIGURE 3.20 shows the fluctuation of body temperature of the antelope ground squirrel and the camel during the course of one day. Notice that the figures for the camel are the same as in previous questions, a maximum body temperature of 40·5°C and a minimum of 34.5°C.

A The antelope ground squirrel starts its daily activity by foraging for food at about 06.00 hours. What happens to its body temperature during the period marked X, and how is this related to the animal's water economy?

B Having dashed about furiously at high speed, perhaps travelling as much as 100 metres from its burrow, the exposure to the burning sun and sand cause the temperature rise we have just noted. The animal then returns to its burrow and flattens itself against the relatively cool floor (period Y in FIGURE 3.20). It then re-emerges and resumes its furious activity, returning again to the burrow when its body temperature has risen above that of the environment (42°C).

i From FIGURE 3.20, what effect has the resting period had on the animal's body temperature?

ii How does resting produce this change in body temperature?

iii What is the significance of the resting period from the point of view of the animal's water balance?

FIGURE 3·19

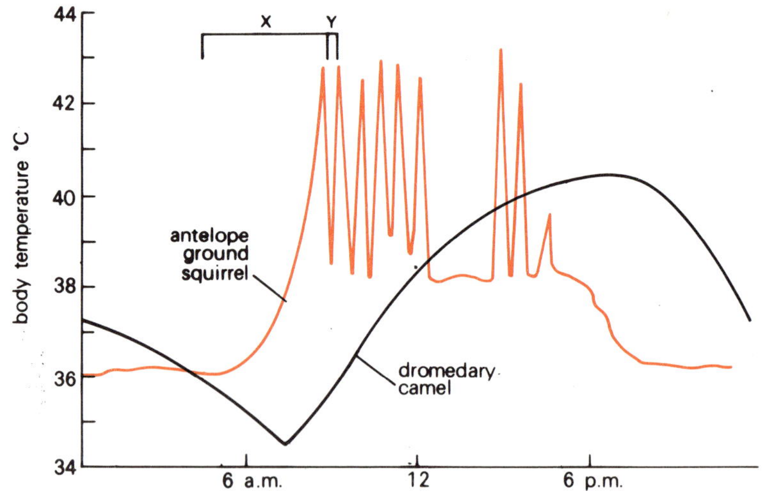

FIGURE 3·20

Finally in this section on temperature control and homeostasis, we will look at another small rodent, the kangaroo rat (*Dipodomys spectabilis*, FIGURE 3.13, p 50) which lives in the Arizona desert. We will compare its habits with those of the antelope ground squirrel and see how it manages to balance its water budget.

It can live in the laboratory permanently without water, eating nothing but dry barley seeds. It is unlikely that in its natural habitat it ever has access to water, even to dew. Thus, as with the antelope ground squirrel, this animal could not use large quantities of water to keep cool by evaporation.

Unlike the antelope ground squirrel, it cannot tolerate a body temperature of more than 38°C, and the daytime air temperature of the Arizona desert rises to about 42°C. It avoids this heat by staying in its burrow all day (where the temperature ranges from 23–31°C) and emerging only at night, when the desert temperature is between 15°C and 23°C. If it were exposed to the heat of the desert day it would quickly die.

A FIGURE 3.13 **is a photograph of a kangaroo rat. What feature indicates that it is a nocturnal animal?**

B **Where does this animal (and the antelope ground squirrel) get its water if it never drinks? Consider the chemical equation for the oxidation of glucose (respiration):**

$$C_6H_{12}O_6 + 6O_2 = 6CO_2 + 6H_2O + \text{energy}$$

Here is a source of water. The atomic weight of carbon is 12, of hydrogen 1 and of oxygen 16. Hence the molecular weight of glucose is 72 + 12 + 96 = 180.

What weight of water will be produced in the animal's body when one gramme of glucose is oxidized in this way?

C **Oxidation of 1 g of fat produces 1·1 g of water and of 1 g of protein, 0·3 g of water. 100 g of perfectly dry barley, on which the kangaroo rat was fed in the laboratory, yield 54 g of water. The animal ate 100 g of dry barley in five weeks; thus the total amount of water available to it in this time was 54 g.**

What prediction could be made about its urine, its faeces and the amount of water lost by evaporation from its body? (N.B. The gerbil, which you probably keep at school, has habits and adaptations to desert life similar to those of the kangaroo rat.)

The retinas of vertebrates contain nerve fibres which lead to the brain in the optic nerve. Light falls on these fibres but they are not sensitive to it. The light-sensitive part of the retina is behind them and consists of a layer of special cells called rods and cones (FIGURE 4.1).

They are complicated cells, made up of several parts. The *outer segment* contains light-sensitive pigments: the chemical change caused in these when light falls on them is converted by the rest of the cell into a nervous impulse.

Another part of the rod and cone cells is called the *myoid*; in the retinas of some vertebrates this can contract, altering the position of the outer segment.

FIGURE 4.2b shows that the rods and cones are not connected directly to the fibre of the optic nerve but that there is a third layer of cells between them, called *bipolar cells*.

We can think of each nerve fibre as being a separate 'private line' to the brain.

A **From FIGURE 4.2b, what is the difference in the way in which rods and cones are connected to the brain?**

B **In some vertebrates (e.g. lizards, squirrels) that are only active in daylight, the retina contains only cones and no rods. Such retinas are insensitive to weak light and are probably useless at night, but they are capable of discriminating fine detail in daylight.**

Bearing in mind FIGURE 4.2b, try to explain how such a retina can discriminate fine detail (e.g. distinguishing between a group of minute white spots and not seeing them as one larger, blurred white spot).

C **The dimmest light stimulus that will cause a response in a cone cell is about the same as that causing a response in a rod cell. Some vertebrates which are active only at night (e.g. bush baby, bats, nocturnal lizards and snakes) have only rods in their retinas.**

i **Why should such retinas not be able to discriminate fine detail?**

ii **Explain why, because the rods are slender and closely packed in vast numbers, the dimmest light to which such**

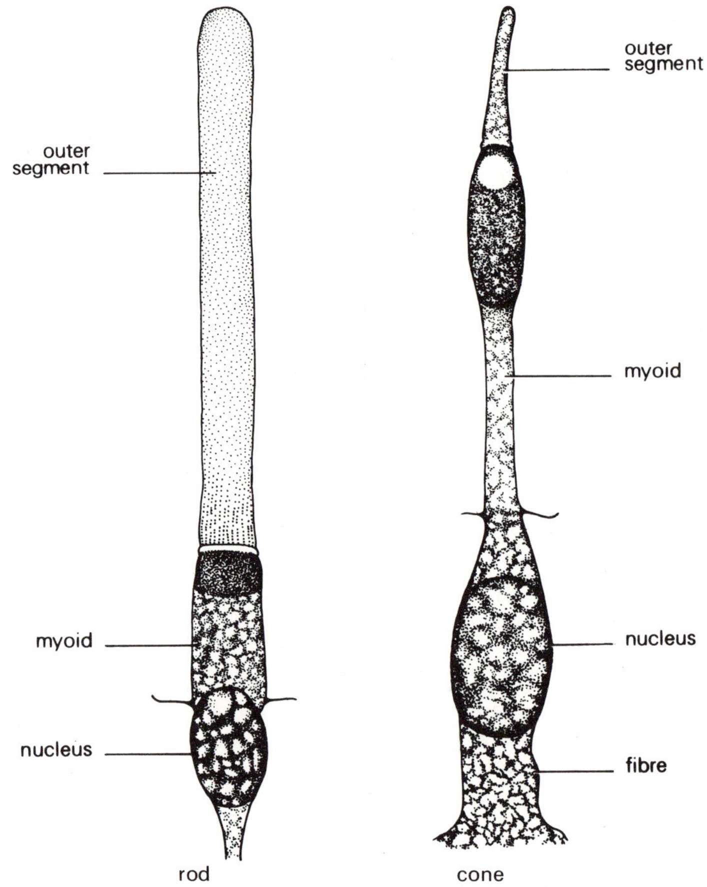

FIGURE 4·1

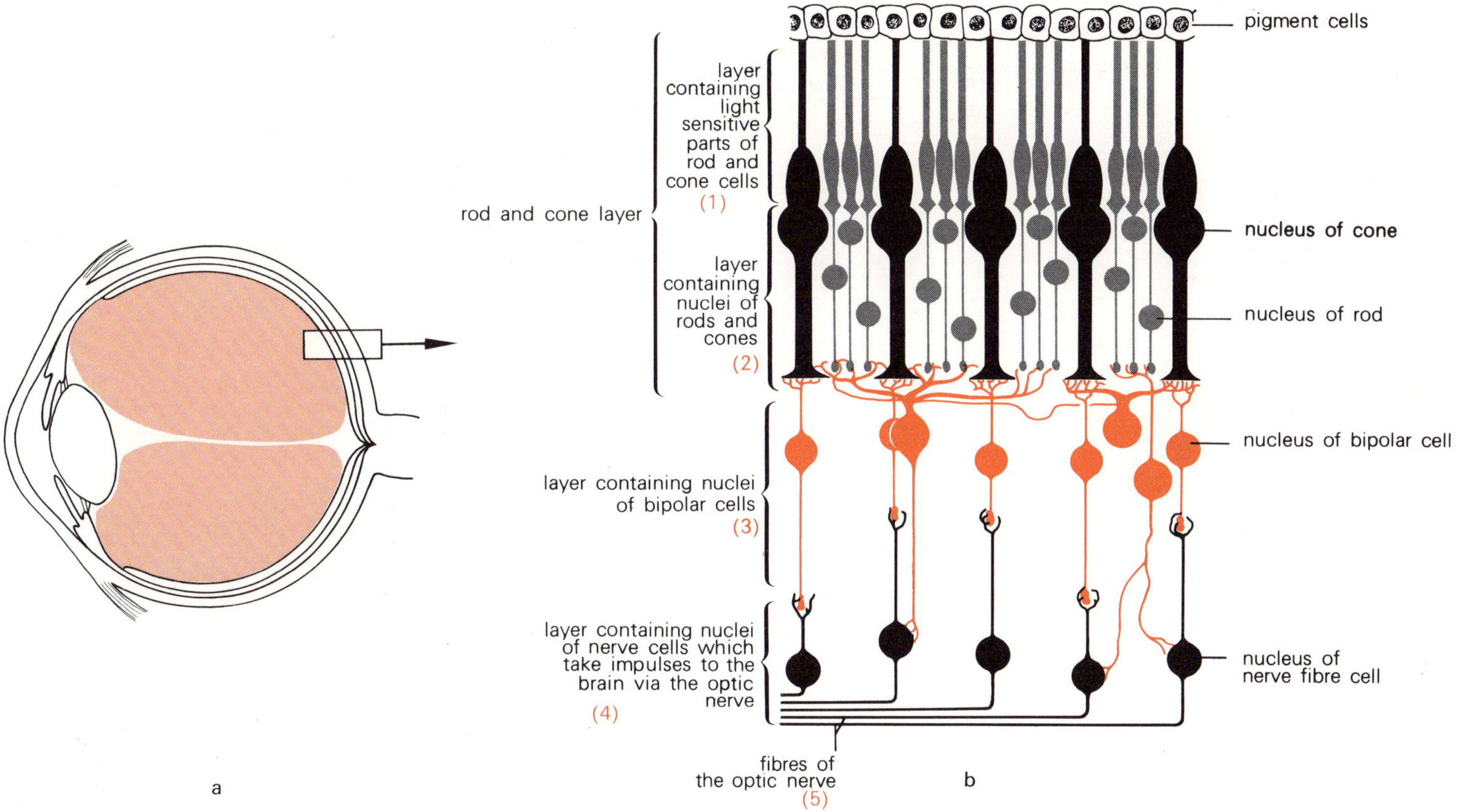

FIGURE 4·2

a rod-retina can respond is much lower than that to which a cone retina can respond. In other words, explain how the rod retina is much more light-sensitive than the cone retina. FIGURE 4.2b will help you to answer this question.

D We see from the previous questions that the discriminating power and the light-sensitivity of cone and rod retinas are not caused by the properties of the individual visual cells themselves. What are they caused by?

(The human retina contains only cones at the *fovea centralis*; outside this small area there are both rods and cones. It is easy to demonstrate that the night vision due to rods can produce only poor discrimination.

Light your bedroom one night with a neon night light and when you have become accommodated and can see clearly, try to read a newspaper.)

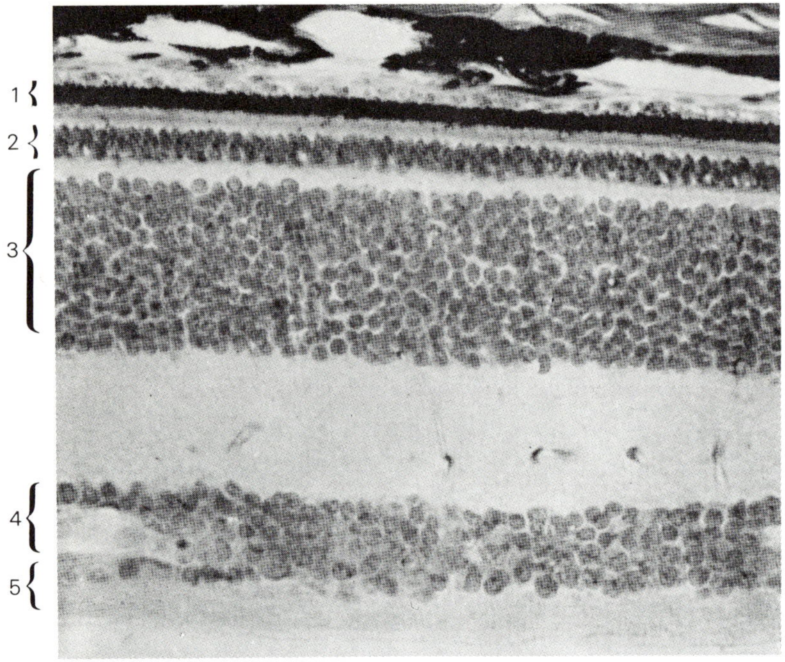

FIGURE 4·3

FIGURES 4.3 and 4.4 are photographs of sections through the retinas of two mammals. The nuclei of the cells show up clearly as black blobs, but the connections between the cells are hazy and difficult to make out.

First examine FIGURE 4.2a carefully and make sure that you understand how the five layers shown are situated. The layers are:

1 The layer containing the light-sensitive parts of the rods and cones.
2 The layer containing the nuclei of the rod and cone cells (one nucleus to each rod and each cone cell).
3 The layer containing the nuclei of the bipolar cells.
4 The layer containing the nuclei of the nerve fibre cells.
5 The layer of nerve fibres leading to the optic nerve.

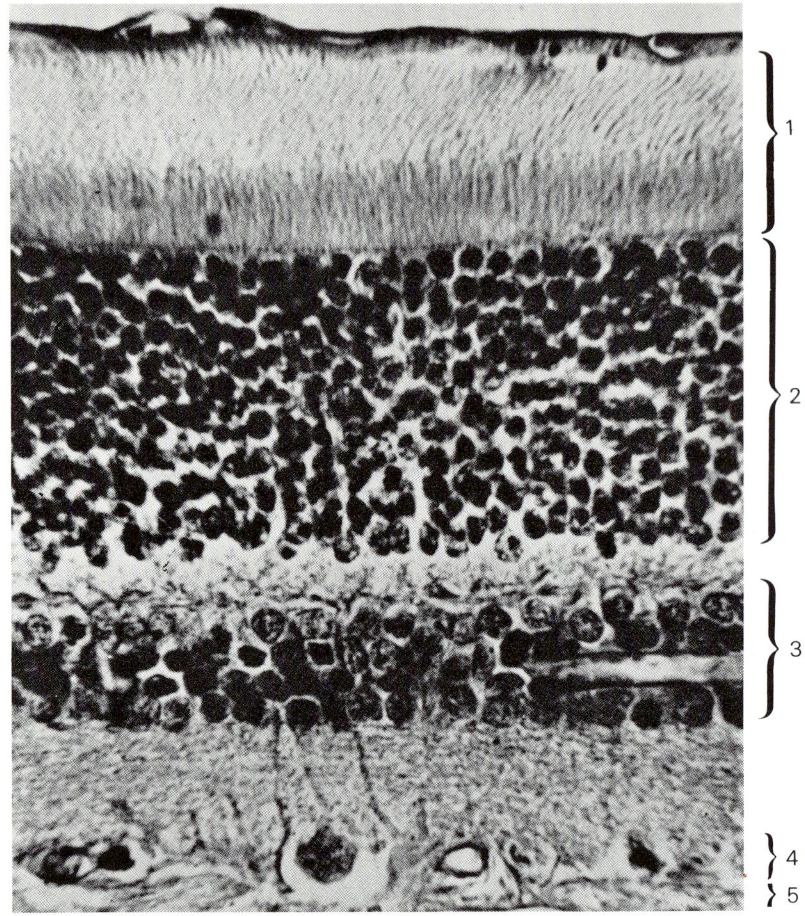

FIGURE 4·4

Identify these layers in FIGURES 4.3 and 4.4.

A How does layer 2 in FIGURE 4.3 compare with layer 2 in FIGURE 4.4 regarding:

i thickness of the layer;

ii number of nuclei in the layer?

B **How does layer 3 in FIGURE 4.3 compare with layer 3 in FIGURE 4.4 with regard to:**

i thickness of the layer;

ii number of nuclei in the layer?

C **How does layer 4 in FIGURE 4.3 compare with layer 4 in FIGURE 4.4 with regard to:**

i thickness of the layer;

ii number of nuclei in the layer?

D **One of these retinas belongs to a nocturnal mammal (i.e. one which is active and uses its eyes only at night), the other to a diurnal mammal (i.e. one which is active and uses its eyes during the day and at night).**

If you remember the answer you gave to question 58 C, the relative thicknesses of layers 2, 3 and 4 will tell you which retina is which.

i Work out which retina is which.

ii Say how the relative thickness of each of the layers 2, 3 and 4 supports your answer.

60

In the mammals that have the highest powers of visual discrimination (the primates, e.g. apes, monkeys and man), there is a special area of the retina called the *fovea centralis*. In man it is a spot 0·5 mm in diameter. It lies on the principal axis of the lens. The object being looked at has its image formed on the fovea, which is specialized for fine discrimination. Thus when you read this page, the images of the words you are reading fall successively on the fovea. Other words (in the lines above and below those you are reading) are not seen clearly; to read them you have to move your eyes so that their image falls in turn on the fovea.

FIGURE 4.5 shows a transverse section of a mammalian fovea. It contains only cones. Other parts of the mammalian retina contain a mixture of rods and cones. In man there are about 30 000 cones in the fovea alone.

A **How is the fact that the fovea contains only cones related to the fact that it is the part of the retina used for the finest discrimination?**

B **Remember how FIGURE 4.5 fits in to the eye (check with**

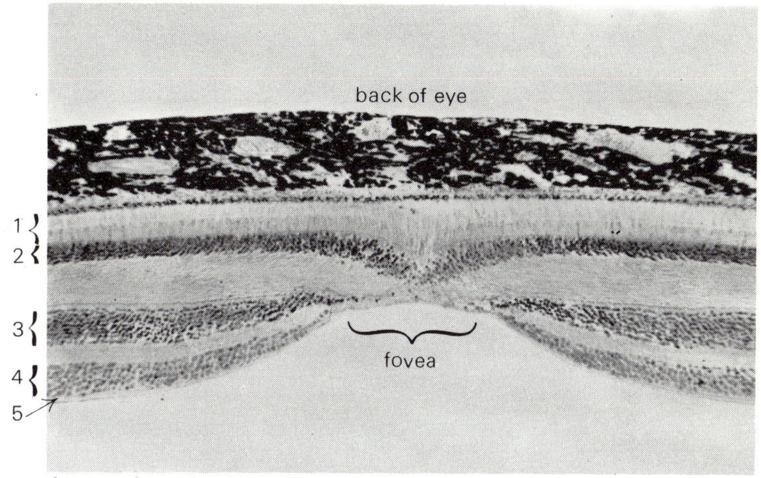

FIGURE 4·5

FIGURE 4.2a). Be sure you know the direction the light is coming from and where the image falls.

Why will the image falling on the rod and cone layer (layer 1) on either side of the fovea not be as sharp as that falling on the cone layer at the fovea?

C **The cones at the fovea have the same connection to the brain as other cones. If this is so, where must the bipolar cells and nerve-fibre nuclei (layers 3 and 4) of the foveal cones lie?**

61

Many vertebrates are capable of being active and therefore of using their eyes at any time of day or night. The retinas of these animals contain both rods and cones (as does the human retina). The rods make the eye sensitive to dim light, the cones give good discriminating power in brighter light.

Some of these animals, for example, cows, sheep, elephants and large carnivorous mammals, have no means of controlling the amount of light that enters the eye; but others do.

A **Under what conditions of illumination will it be an ad-**

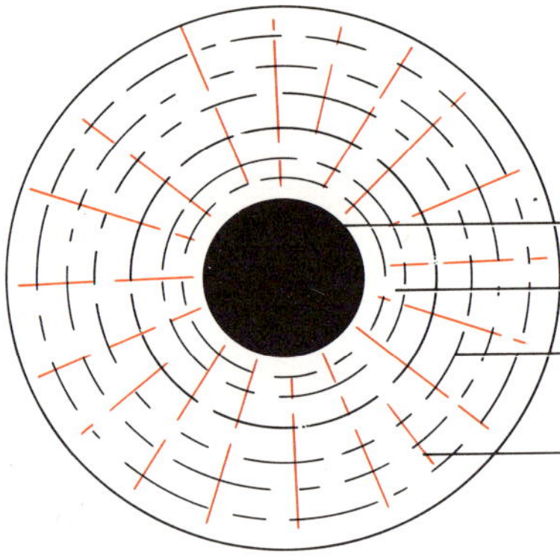

FIGURE 4·6

vantage to admit as much light to the eye as possible? Give a reason for your answer.

B **Under what conditions of illumination will it be an advantage to cut down the amount of light entering the eye as much as possible? Give a reason for your answer.**

C **The iris in man contains two sets of muscle fibres, one arranged in a circular fashion, the other running like lines of longitude from the pupil to the edge of the iris (FIGURE 4.6).**

The only active thing these muscles can do, like all muscles, is to contract.

i **What effect will contraction of the circular muscle fibres have on the size of the pupil?**

ii **What effect will the contraction of the longitudinal muscle fibres have on the size of the pupil?**

D **Each of these sets of muscles receives a separate nerve supply causing it to contract. Contraction is a reflex action and cannot be controlled by an effort of will. Reflex actions are set off by the stimulation of a sensory cell. A simple reflex arc contains four cells, sensory cell→sensory nerve→ motor nerve→motor cell (e.g. muscle cell).**

Where would you think the light-sensitive sensory cells that set off the reflex movements of the iris are likely to be situated?

E **Some nocturnal animals, such as cats and snakes, which like also to bask in the sun, have pupils that close to a vertical slit. Why would this provide better protection for a sensitive retina than the type of iris closure found in man?**

62

This question is concerned with the way in which the human eye is able to form sharp images on the retina of objects which may be as close to it as 25 cm or as far away as several hundred metres.

First look at FIGURE 4.7. It shows in a simplified way how a bi-convex lens, such as the lens found in the eye, forms an image of a *distant* point.

When light enters glass from air it is bent. Any ray of light that is perpendicular to the lens, such as the one marked A, is bent so that it passes through a point on the far side of the lens called the *principal focus*. The distance from the centre of the lens to the principal focus is called the *focal length* of the lens. Only four rays of light have been drawn in FIGURE 4.7 in order to make the diagram simple and clear. If you focus an image of the sun on a piece of paper with a magnifying glass all the rays are bent so as to meet at the principal focus of the lens.

But the place where the image is formed is *not fixed*; it depends on the position of the object. FIGURE 4.7 shows the image at a distance from the lens equal to the focal length of the lens; this is the case when the object is a long way from the lens.

FIGURES 4.8a and b show two more positions of the object, both closer to the lens. In FIGURE 4.8b the object is only a little further from the lens than twice its focal length. Here only two rays of light have been drawn so as to make the diagram as simple as possible. They show how an image of the tip of the object arrow is formed.

A **What has happened to the position of the image in FIGURES 4.8a and 4.8b compared with its position in FIGURE 4.7? (Measure its distance from the centre of the lens.)**

B FIGURES 4.8a, b and c **can be thought of as representing the lens and retina of the eye. The distance from the lens to the retina cannot be changed in a human being. In FIGURE 4.8a the lens has bent the light from the arrow tip sufficiently to form a sharp image on the retina.**

Where would a sharp image be formed in FIGURE 4.8b ?

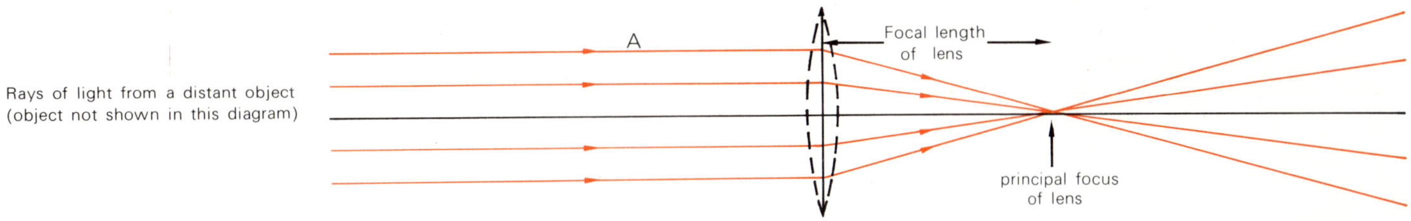

FIGURE 4·7

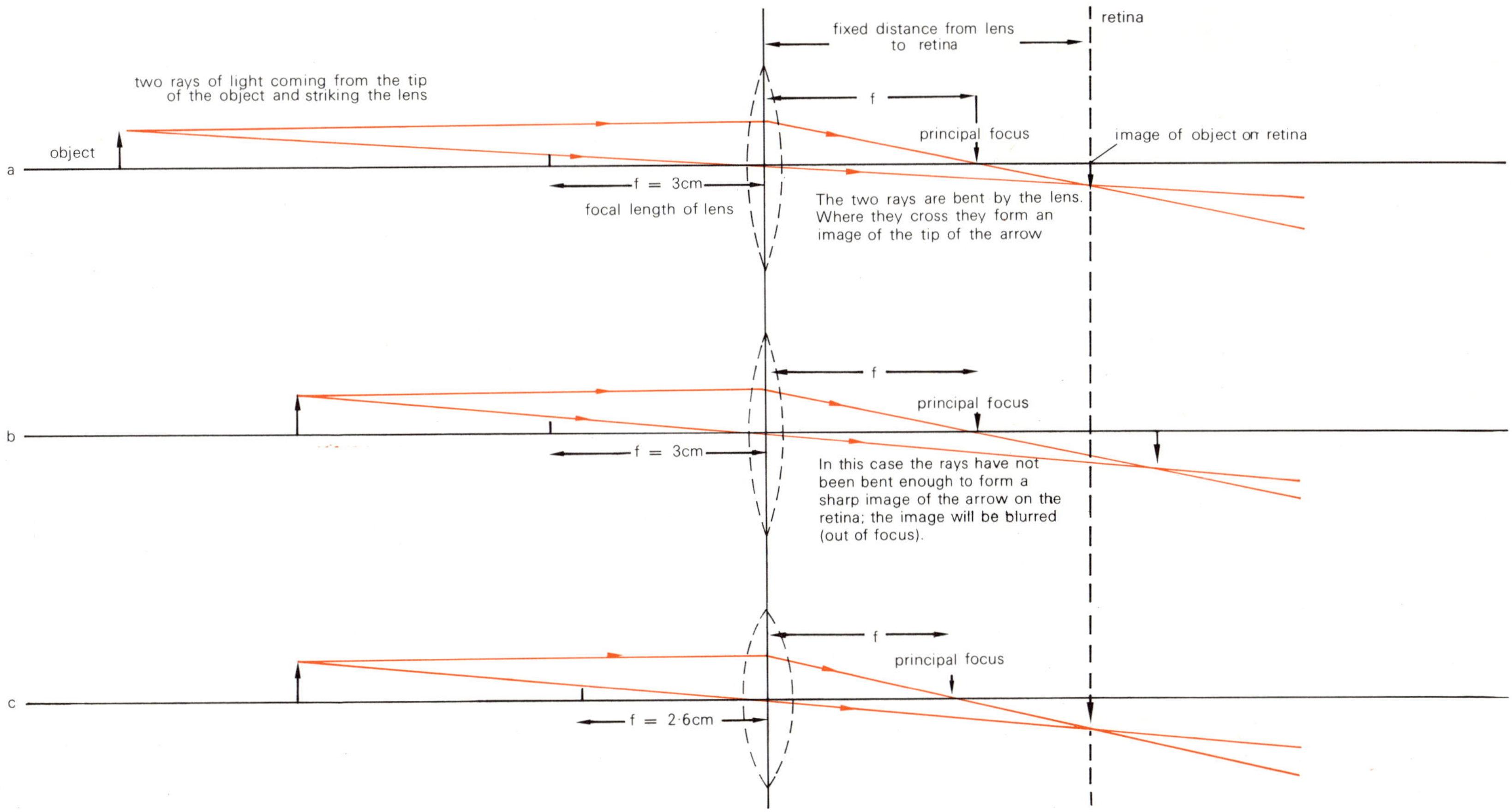

FIGURE 4·8a, b & c

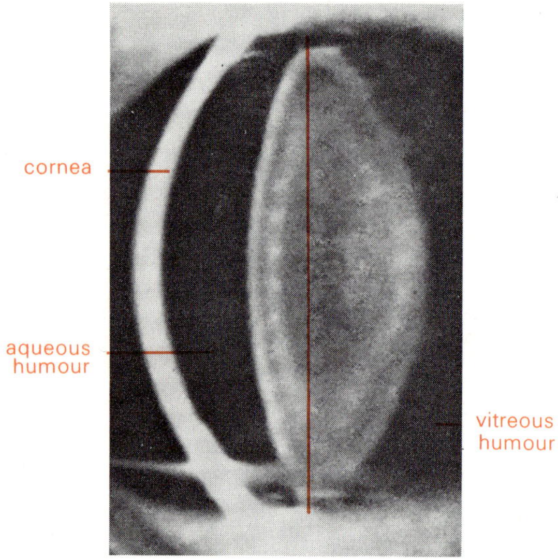

FIGURE 4·9a

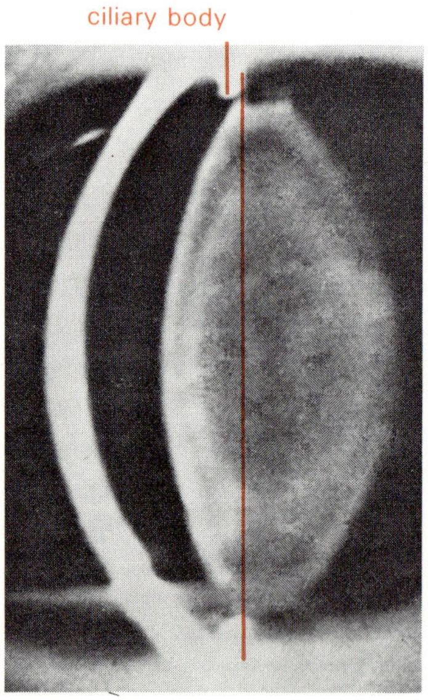

FIGURE 4·9b

63

Thus it seems that when an object is close to the eye a sharp image of it cannot be formed on the retina. It would be formed behind the retina, but that, of course, is not possible; instead a blurred image is formed on the retina. Hold this page about 10 cm from your eye and try to read it; you will see then what a blurred image looks like.

In fact, however, the human eye is able to see objects clearly when they are as close to it as 25 cm. It cannot do this by changing the fixed distance between the lens and the retina, but it can by changing the shape of the lens.

This change is shown in FIGURE 4.9, two photographs (side views) of the human lens. In a the eye was looking at a distant object; in b it was looking at a close object.

A **How did the thickness of the centre of the lens change when the eye looked at the closer object (measure the widths)?**

B **Take a piece of tracing paper and trace the shape of the lens in** FIGURE 4.9a. **Then place the paper over** FIGURE 4.9b **and trace the second shape of the lens. The line drawn on the photographs will help you; it may also help to use a pencil for one lens and ballpoint pen for the other.**

Describe the change in shape of the front and the back surfaces of the lens when the eye looks at an object close to it.

C FIGURE 4.8c **shows a fatter lens than** FIGURE 4.8a **and** 4.8b, **like the lens in** FIGURE 4.9b.

i **What is the focal length of this new lens?**

ii **How does the way it bends the upper ray of light compare with the way in which this ray is bent by the lens in** FIGURE 4.8b **?**

iii **The object in** FIGURE 4.8c **is the same distance from the lens as that in** FIGURE 4.8b. **The image in** FIGURE 4.8b **would have been formed in focus behind the retina (if that were possible).**

Where is the image formed in focus in FIGURE 4.8c **?**

64

In the group of fish which have gristle skeletons (such as sharks and skates), there is a different mechanism for keeping objects at different distances from the lens in focus. Here there is a muscle inside the eye which is able to pull the lens forward, away from the retina. The shape and focal length of the lens are not changed. (This is the mechanism which is used for focussing in expensive cameras.)

Explain how this will allow both distant and near objects to be focussed sharply on the retina. FIGURES 4.8a **and** b **will help you to answer this question.**

65

(In FIGURE 4.10, the bending of light due to the cornea cannot be shown and the light rays have been drawn incorrectly passing straight through the cornea to the lens unbent.)
FIGURE 4.10a shows a normal eye. An image of a distant object is formed in focus on the retina, and the distance between the lens and the retina equals the focal length of the lens (compare this case with FIGURE 4.8a, where an object close to the eye has its image focussed on the retina, the retina being beyond f ; f can change but the distance from lens to retina cannot).

A **Suppose this were the same eye as the one shown in** FIGURE 4.8a; **would you expect the lens to be thinner or fatter than the one shown in** FIGURE 4.8a **? Give your reason.**

B **In** FIGURE 4.10a, **the distance between the lens and the retina is correctly suited to the eye.** FIGURE 4.10b **shows an abnormal eye in which the eyeball is too short for the lens. When the lens is at its thinnest, rays of light cannot be brought into focus on the retina but would be brought to focus behind it, if that were possible. As a result, the eye keeps the lens fatter even when it is looking at a distant object.**

Explain why, if the lens is fatter, the image of a distant object will be formed in focus on the retina.

C **What effect do you think this condition will have on the eye's ability to form an image of an object close to it in focus on the retina?**

D **Explain how the spectacle lens shown in** FIGURE 4.10c **corrects this defect and enables an image of a close object to be formed in focus on the retina.**

E FIGURE 4.10d **shows an abnormal eye in which the eyeball is too long for the lens. The lens is shown at its thinnest and the eye is looking at a distant object.**

Explain why the distant object will not be seen clearly.

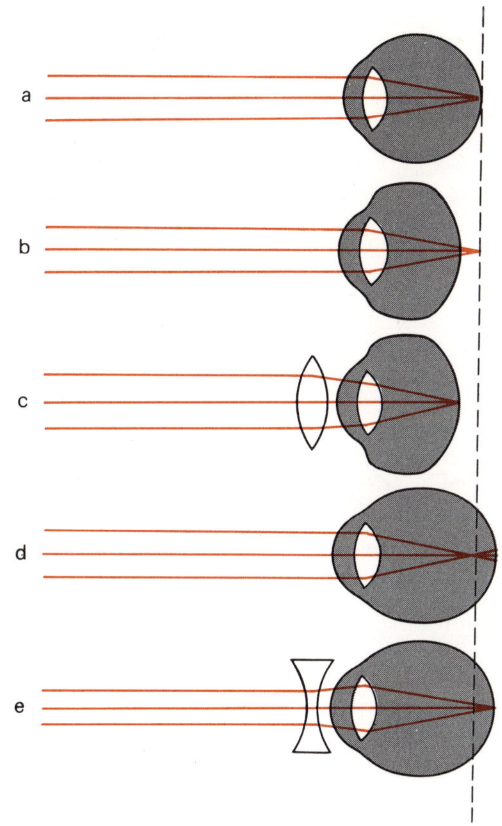

FIGURE 4·10a, b, c, d & e

F **A person with this defect can see clearly objects very close to him. Look back to** FIGURES 4.8b **and** c. **Explain why the lens in this eye would not have to become so fat as that shown in** FIGURE 4.8c **in order to bring objects close to the eye into focus on the retina.**

G **Explain how the spectacle lens shown in** FIGURE 4.10e **corrects this defect and enables an image of a distant object to be formed in focus on the retina.**

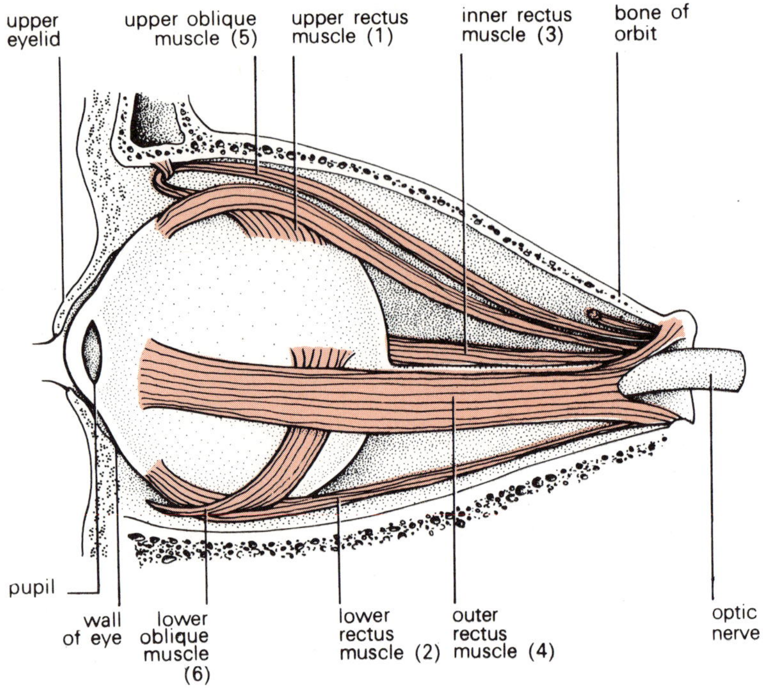

FIGURE 4·11

66

FIGURE 4.11 is a stereogram of the human eye lying in its bony socket (orbit) in the skull. At the front the eyelids can be seen, at the back, the optic nerve leading to the brain.

A **What is the name of the white wall of the eye which we can see in the picture?**

B **The drawing shows the six muscles that move the eye. Each muscle is attached at one end to the wall of the eye and at the other to the bone of the orbit. The combined action of these muscles makes possible a wide and detailed range of eye movement. This question asks you to consider the movement produced by each of the six muscles.**

Look first at the four rectus **muscles, labelled** 1, 2, 3 **and** 4. **They are equally spaced and are attached to the top, bottom and sides of the eye. What eye movements would you expect as a result of the contraction of the following muscles** alone **?**

- **i** **upper rectus muscle**
- **ii** **lower rectus muscle**
- **iii** **inner rectus muscle**
- **iv** **outer rectus muscle.**

C **Muscles** 1, 2, 3 **and** 4 **are attached to the bone of the orbit. Of the remaining two (the upper and lower oblique muscles, numbers** 5 **and** 6 **in the drawing), the upper oblique is also attached to the back of the orbit and it takes a sharp turn through a bony pulley before being attached to the wall of the eye. The lower oblique muscle** (6) **is attached to the front of the orbit.**

What eye movement would you expect if the following muscles contracted alone? In order to answer the question, imagine the position of the muscle in your own left eye and think of the movement it would produce in your own eye.

- **i** **upper oblique muscle.**
- **ii** **lower oblique muscle.**

67

In question 59 (p 58) we saw some of the characteristics of the retina of a nocturnal mammal (bush baby). Other parts of the nocturnal eye, indeed the whole eye, show special adaptations to

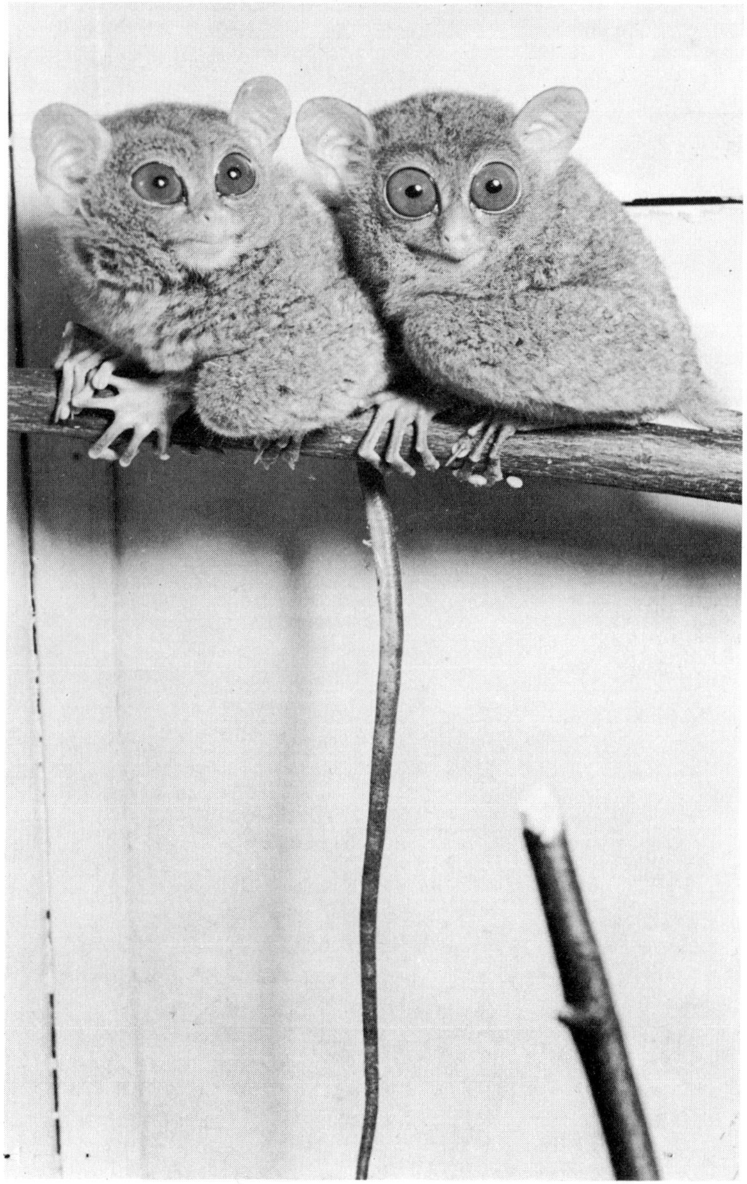

FIGURE 4·12

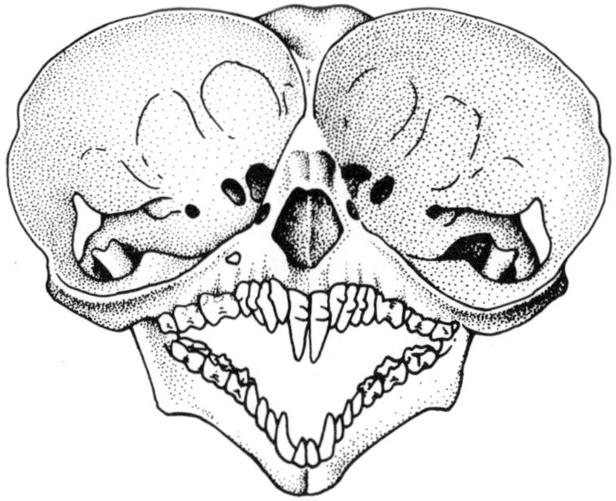

FIGURE 4·13

vision in conditions of dim light. One of the features is well illustrated in FIGURE 4.12, a photograph of two Tarsiers, and FIGURE 4.13, a drawing of a front view of a Tarsier skull.

Tarsiers belong to the same group of mammals (the Primates) as man, though they are primitive members. They are confined to the Phillipine Islands. They are about 15 cm long and live exclusively in trees in the rain forest. They sleep during the day, clinging to the bark of a tree. Their hands are well adapted for clinging, and their limbs for leaping. They can jump as much as two metres at a time, catching a flying insect, on which they feed, in mid leap and land safely on another branch. Thus they have acute vision, though the retina contains only rods and no cones. When leaping the animal closes its eyes, so it must have calculated the position of its prey and of its landing stage accurately before making the leap.

A **From** FIGURES 4.12 **and** 4.13**, what is the most striking feature of the eyes of** *Tarsius* **?**

B **How will this fact contribute to the efficiency of the Tarsier eye?**

C **How is the fact that the retina of the Tarsier contains only rods and no cones related to its nocturnal way of life?**

D **The eyes of the Tarsier are not spherical like human eyes (**FIGURE 4.11**), but are slightly tubular.** FIGURE 4.14 **is a drawing of a section through a Tarsier eye, showing that it is longer than it is wide.**

We have seen that the light-admitting front part of the eye is very big; if the whole eye were spherical, as is the human eye, and this part had its large size, the whole eye would be so big that the skull could not contain it.

FIGURE 4.13 **shows that the orbits are already as large as they possibly could be. Thus the tubular shape enables the eye to have a larger light-admitting front part relative to the size of the whole eye than does a spherical shape.**

However, the tubular shape has its drawbacks. Look back to question 66 **and suggest one obvious disadvantage of a tubular eye compared with a spherical one.**

E **Comment on the fact that the Tarsier has an extremely mobile neck and can turn its head through 180°, so that its eyes face backwards.**

F **The large and tubular eyes and mobile neck that we have seen in** Tarsius **are found in many other nocturnal vertebrates, though all these features may not be present at one time.**

Name one common nocturnal animal which has large eyes and a very mobile neck.

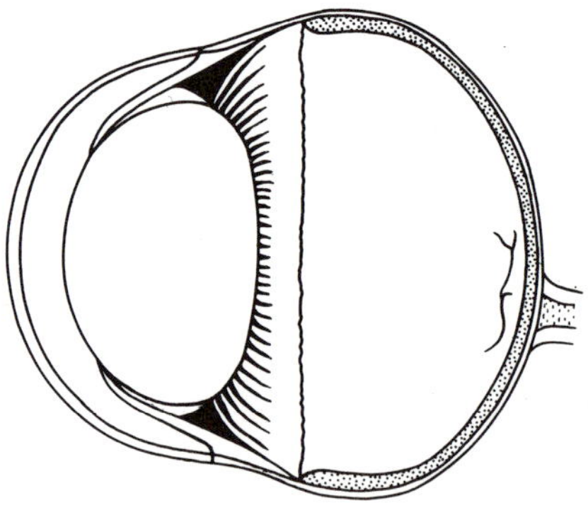

FIGURE 4·14

5 Nutrition

FIGURE 5.1 is a drawing of a man's head cut vertically through the middle and looked at from the side (a sagittal section).

68

A **What are the structures labelled** D, E, F **and** G **?**

B **What is the space labelled** A **? Name any two special structures it contains and say what their function is.**

C **i** **What are the structures labelled** B **and** C **?**

ii **Apart from the crocodile (a reptile), mammals are the only vertebrates that possess these structures. What special ability do they confer on the animal from the point of view of digestion of food?**

69

Mammals vary in the amount of chewing they give to their food. Some slice their food and bolt it fairly quickly (e.g. dog); some chew it for a very long time (e.g. cow); others are in between (e.g. man).

State two advantages connected with the digestive juices of chewing food before swallowing it.

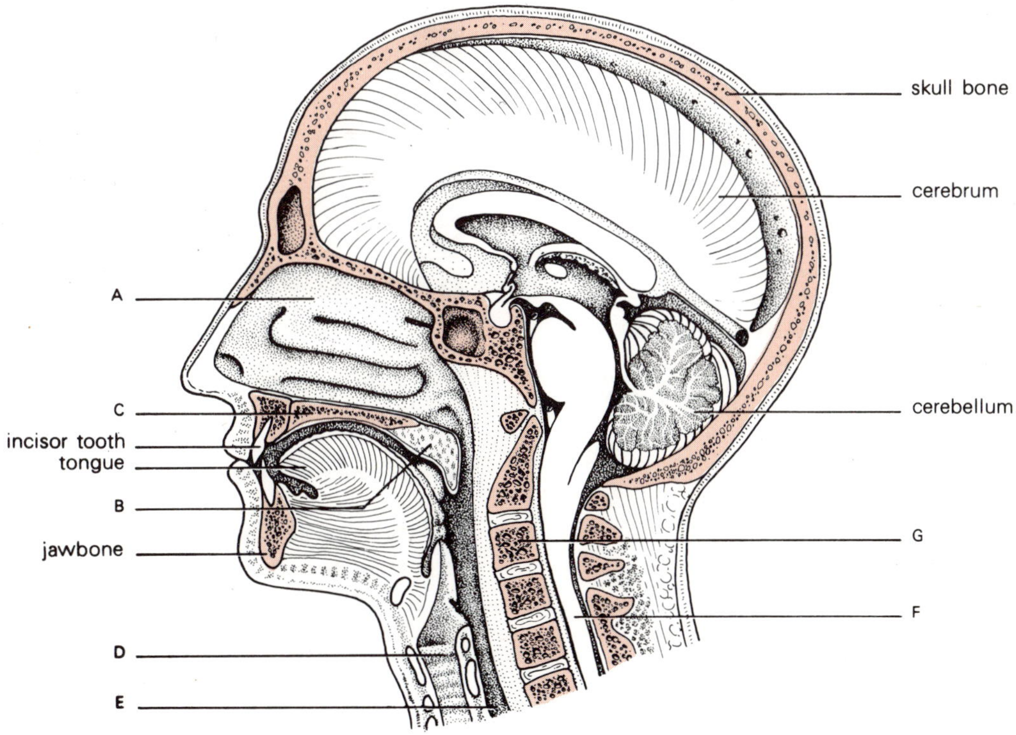

FIGURE 5·1

70

FIGURE 5.2 is a side view of a human skull.

A **What are the structures labelled** A **and** B **?**

B **Each of the teeth on the left side of the lower jaw has been numbered. Which are the molars, the premolars, the canine and the incisors?**

71

FIGURE 5.3 is a drawing of the skull of a dog. The incisors, canines, premolars and molars have been labelled. It is not possible to show the three-dimensional shape of the teeth in this figure, or to show the kind of movement of the lower jaw that the jaw joint allows. Check on these features in the dog skull in your school laboratory and then answer the following questions:

A **How do the numbers of teeth in the upper and in the lower jaws of the dog compare with the numbers in the upper and lower jaws of a human adult?**

B **If the premolars and molars had not been labelled in** FIGURE 5.3, **and you did not find out the answer from anyone else or from a book, how could you tell the difference between the premolar and molar teeth?**

C **What are the functions of the incisor and canine teeth of a dog, and how is their shape suited for these purposes?**

D **How does the movement of the lower jaw of the dog differ from that of a human being (try moving your own)?**

E **How are the two kinds of jaw movement related to the feeding habits of the dog and the human?**

72

FIGURE 5.4 is a drawing of a sheep skull.

What are the names and functions of the parts labelled A, B, C **and** D **?**

73

FIGURE 5.5 shows

a a human molar tooth in situ;

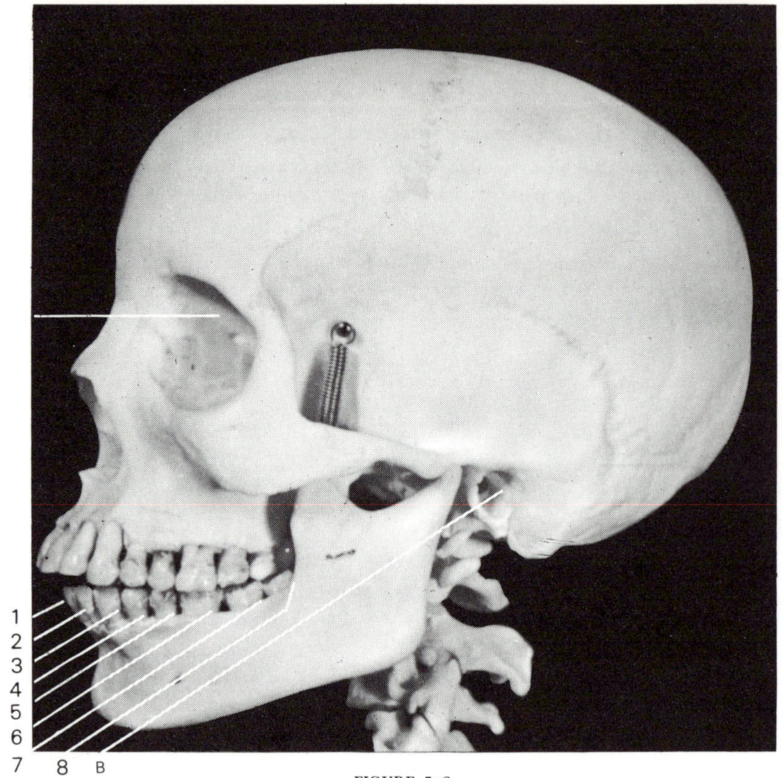

FIGURE 5·2

b a sheep molar tooth newly erupted;

c a sheep molar tooth after wear.

A **Name two structures contained in the pulp of the tooth.**

B **Give two reasons why it is important for the enamel of teeth to be a hard substance.**

C **How does the outside of the sheep molar differ from that of the human molar?**

D **Apart from wearing, how does the ridging of the sheep molar differ from that of the human molar?**

E **i** **Describe the arrangement of the exposed cement, enamel and dentine in the worn sheep molar.**

ii **Why does wear produce this pattern in the sheep molar?**

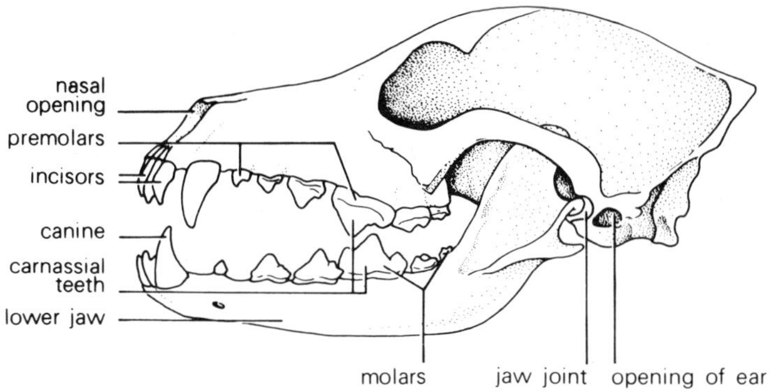

FIGURE 5·3

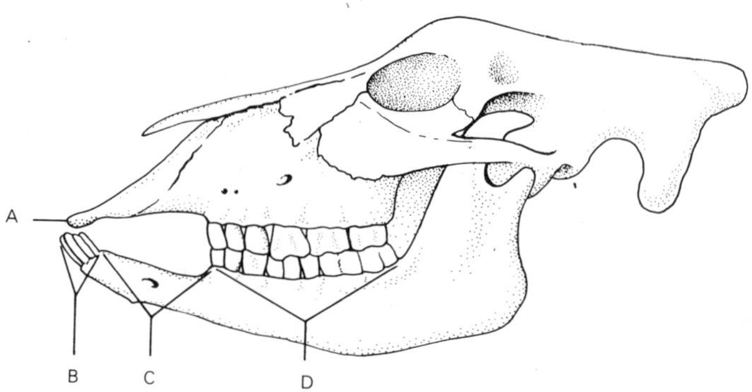

FIGURE 5·4

iii How is this worn pattern an advantage to the adult sheep compared with the unworn pattern of the newly-erupted tooth?

F How does the root of the sheep tooth differ from that of the human tooth, and what is the significance of the difference?

G Mammals have two sets of teeth, a milk set and a permanent set. The change over is gradual. In a human being the milk set (20 teeth) is complete by the age of two; between the ages of six and twelve all the milk teeth are replaced by 32 permanent teeth; at about the age of twelve 8 molars appear; and sometime after the age of seventeen, the last molars (wisdom teeth).

Why do you think it is an advantage for mammals to have two sets of teeth rather than one?

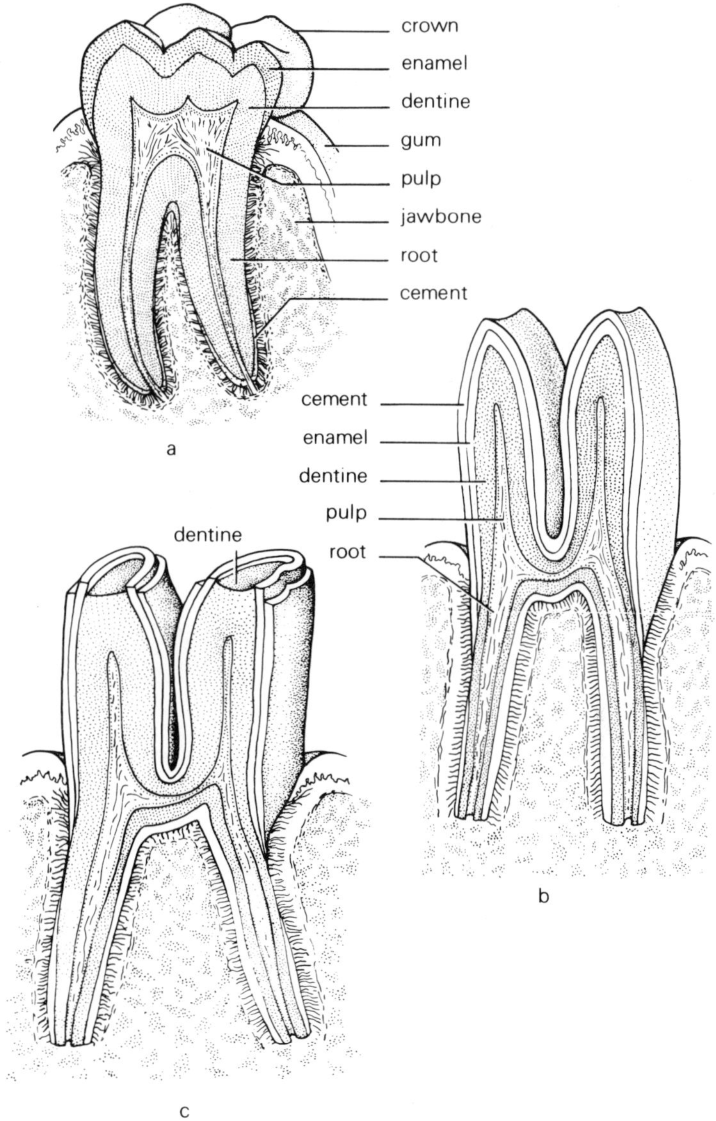

FIGURE 5·5a, b & c

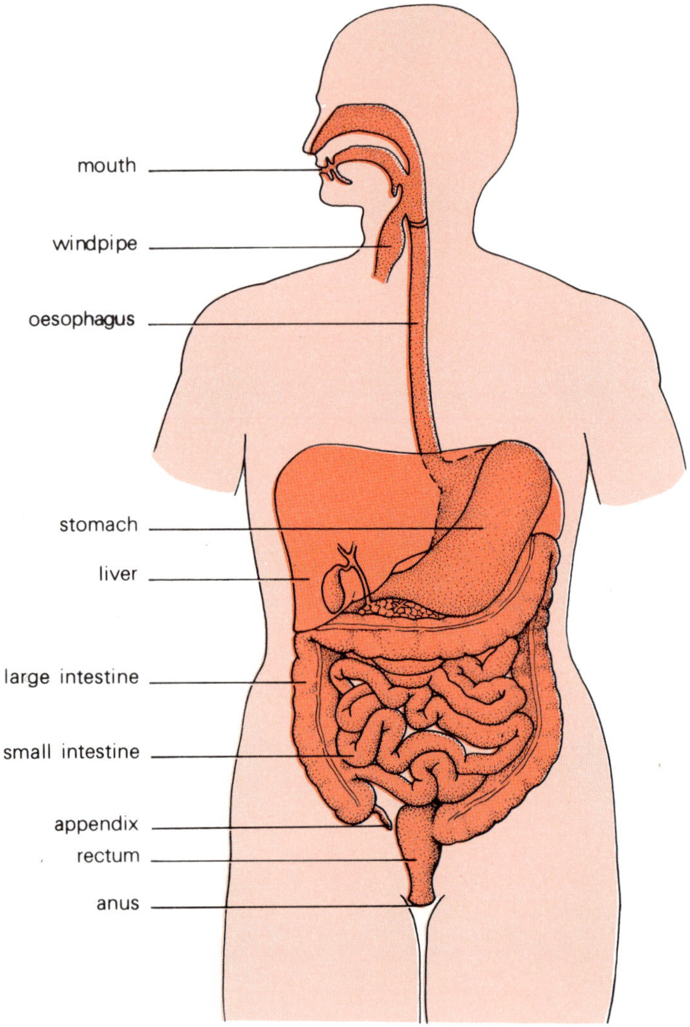

FIGURE 5·6

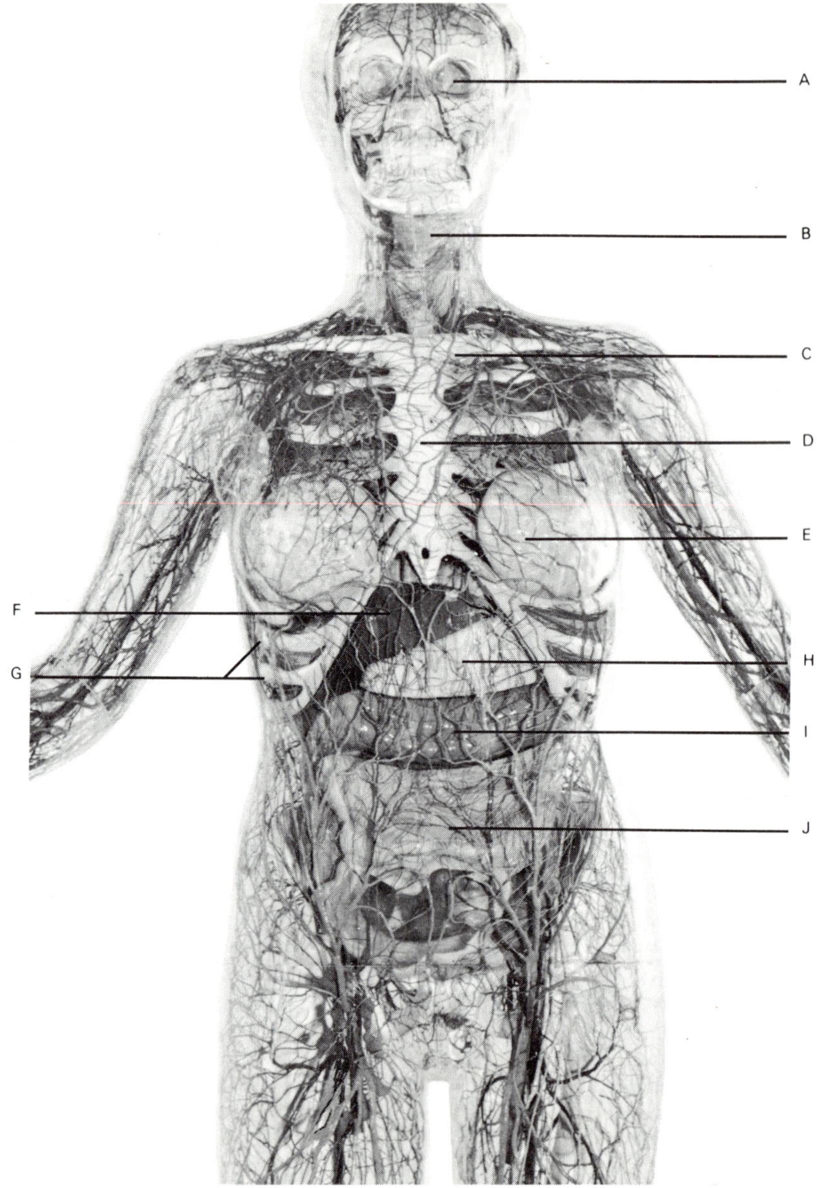

FIGURE 5·7

74

FIGURE 5.6 is a drawing showing the positions of some of the parts of the human gut and the liver. FIGURE 5.7 is a photograph of a life-sized model of a woman in plastic. The skin and muscles are represented by transparent plastic, so the organs underlying them can be seen.

A **What are the structures labelled** A **to** J **?**

B **What glands associated with the gut are not shown in** FIGURES 5.6 **and** 5.7 **?**

C FIGURE 5.8 **shows a bolus of food moving along the oesophagus.**

- **i** **Describe how the activity of the circular muscle (part of which is shown) brings about this movement.**
- **ii** **In which direction will the bolus of food shown move?**

D **This layer of circular muscle runs the whole length of the gut, though it is thicker in some parts than in others. In three places it is particularly thick and forms valves which, until they are relaxed, prevent or at least hinder the gut contents from moving on.**

What are the names of these valves, and where are they situated?

75

Which of the following statements is the correct description of the process of digestion?

- **a** **the churning of food in the stomach**
- **b** **the passage of food along the gut of an animal**
- **c** **the conversion of larger, insoluble food molecules into smaller, soluble ones**
- **d** **the separation in the gut of useful from the useless food substances.**

76

Fats, proteins and starches are insoluble. In the gut they are converted into soluble materials which can be absorbed into the blood stream. One molecule of each of these substances consists of several (sometimes hundreds) of smaller molecules linked together. Their chemical conversion into soluble materials consists in unlinking the smaller molecules from one another, something

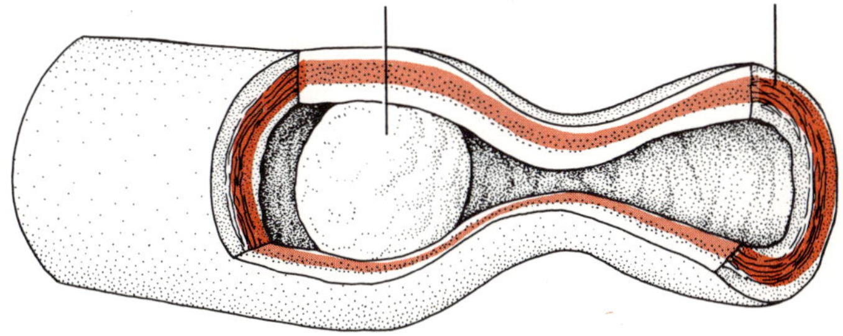

FIGURE 5·8

like unlinking the component trucks of a long goods train. Although fats, proteins and starches are very different chemical compounds, the unlinking process is the same sort of chemical reaction for all of them.

A **What, in simple terms, does the unlinking process consist of?**

B **What is the general name of the catalysts which bring about these unlinkings?**

C **What are the simpler, soluble chemicals that result from the unlinking of:**

- **i** **fats,**
- **ii** **proteins,**
- **iii** **starches?**

77

Liquid is added to the food in the mouth, in the stomach and in the small intestine.

A **What substances does each of these liquids contain which affect the digestion of fats, proteins and starches?**

B **How does the nature of the liquid added in the stomach serve an important function in keeping the mammal free from infection?**

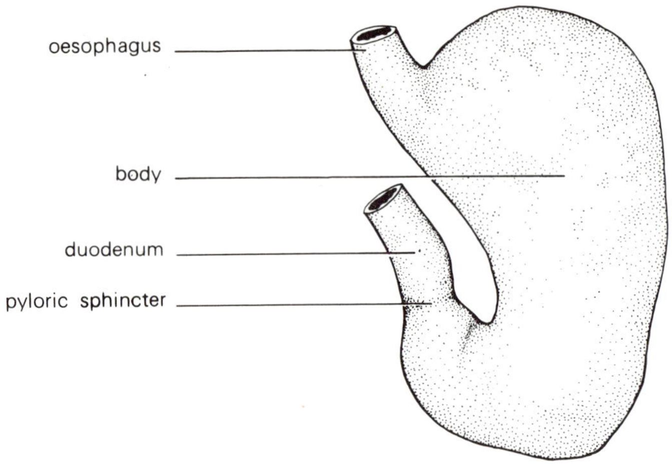

FIGURE 5·9

78

We will look at the pyloric sphincter, which divides the stomach from the duodenum (FIGURE 5.9).

It was thought for a long time that this sphincter was usually tightly closed. Food was thought to be churned in the stomach until it was pulped into chyme, when the sphincter relaxed, allowing the chyme to pass on.

It is now known that only mild movements occur in the upper two thirds of the stomach. Here food is mixed with gastric juice, moistened and softened. Vigorous churning occurs only in the pyloric part of the stomach. Furthermore the pyloric sphincter, though it narrows the exit from the stomach is rarely, if ever, closed.

We must remember that the duodenum is full of liquid material and that peristalsis is taking place in it.

A **If the pyloric sphincter is not closed, what will prevent chyme from passing from the pyloric part of the stomach into the duodenum?**

B **What two useful purposes will the narrowing of the gut at the pyloric sphincter serve regarding**

i **large food particles in the stomach**

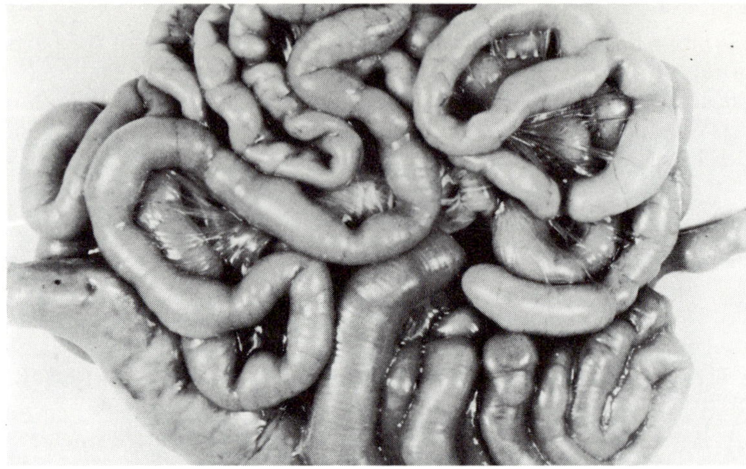

FIGURE 5·10a

FIGURE 5·10b

ii **the contents of the duodenum?**

C **Waves of peristalsis pass along the human stomach at the rate of about three per minute. How long food remains in the stomach depends on how large it is. It can stay for as long as four hours, but some material, particularly liquid**

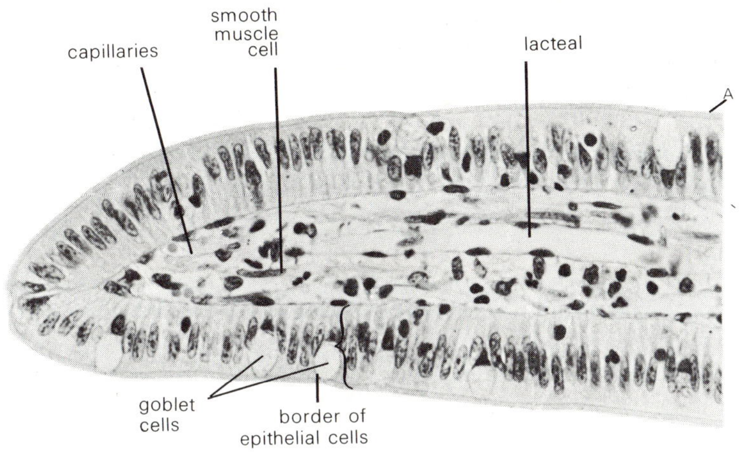

FIGURE 5·10c

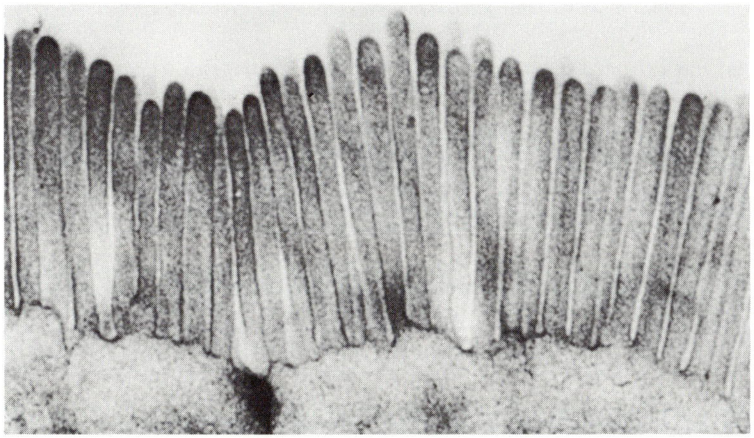

FIGURE 5·10d

food, passes on to the duodenum immediately.

Food always leaves the stomach in small spurts, never in one large mass. Why is this an advantage from the point of view of further digestion in the duodenum and small intestine?

D **By the time the food reaches the small intestine quite a lot of water has been added to it – about 90% of the intestinal contents are water.**

i **What advantage connected with digestion is there in having a large proportion of water in this part of the gut?**

ii **The faeces contain a much lower proportion of water than this. How is the reduction in water content achieved?**

79

FIGURE 5.10a is a photograph of the small intestine of a four week old pig.

FIGURE 5.10b is a photograph (× 14) of a portion of the inner surface of the pig's intestine, showing the villi.

FIGURE 5.10c is a section through the centre of one villus.

FIGURE 5.10d is an electron micrograph × 30 000 of the edge of part of two of the epithelial cells of a villus (e.g. in the region A in FIGURE 5.10c).

A **How do these pictures show that the surface area of the inside of the intestine is very much greater than the surface area of the outside?**

B i **How does the large internal surface area of the intestine contribute to its efficiency in absorbing dissolved foods from the liquid it contains?**

ii **What must happen to the dissolved food material when it arrives inside the villus in order for this efficient absorption to be continued?**

C **Explain the part played in the absorption of food by the following three structures, which can be seen in** FIGURE 5.10c **:**

i **muscle cells**

iii **capillaries**

iii **the lacteal.**

D **Measure the length and width of one villus in** FIGURE 5.10b **and work out its real length and width.**

E **Measure the length and width of one microvillus in** FIGURE 5.10d **and work out its real length and width.**

F **What is the function of the goblet cells seen in** FIGURE 5.10c **?**

80

Some bromine (a dark brown liquid which evaporates readily at room temperature) was placed in a gas jar. The lid was put on and the jar shaken so that the dark brown vapour spread evenly throughout the jar. A second empty jar was placed upside down on the lid of the first, and the lid was gently removed. The jars were not jolted or moved in any way.

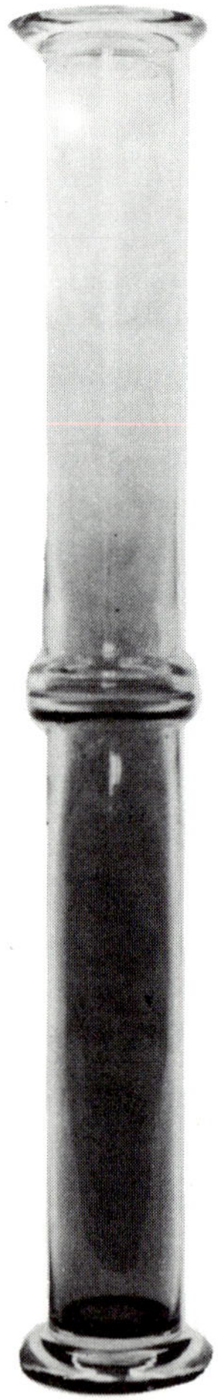

FIGURE 5·11

At the moment of removing the lid from between the two jars, the space inside the bottom jar was a region of high concentration of bromine vapour, the space inside the top jar a region containing no bromine vapour.

FIGURE 5.11 shows the appearance of the jars five minutes later.

A What has happened to the bromine vapour in the five minutes after removal of the lid?

B Bromine vapour is much denser than air. What, then, has caused this change?

C What is the name given to this process, and when will it cease in this experiment?

D The process demonstrated here with bromine is directly related to the uptake of dissolved foods in the intestine. In the intestine, what corresponds to:

i the space inside the lower jar;

ii the space inside the upper jar?

81

An experiment was performed with cat intestine to see whether equal concentrations of three similar simple sugars, glucose, fructose and galactose, were absorbed through the wall into the blood at the same rate. Each sugar was tested separately, but we will assume that apart from using different sugars on different occasions, everything else was the same in each experiment.

The result was that glucose was absorbed twice as fast as fructose and three times as fast as galactose.

In a second experiment the separate sugar solutions were placed inside cellophane tubes surrounded by water, and the rate of diffusion through the cellophane into the water was measured. The three sugars diffused through at the same rate.

A What do these observations tell you about the properties of the cells which line the inside of the intestine?

B When the cells lining the villi are examined under the electron microscope (FIGURE 5.12), it is seen that they contain large numbers of mitochondria, especially towards the outside of the cell. How might this observation be correlated with the experimental evidence presented in this question?

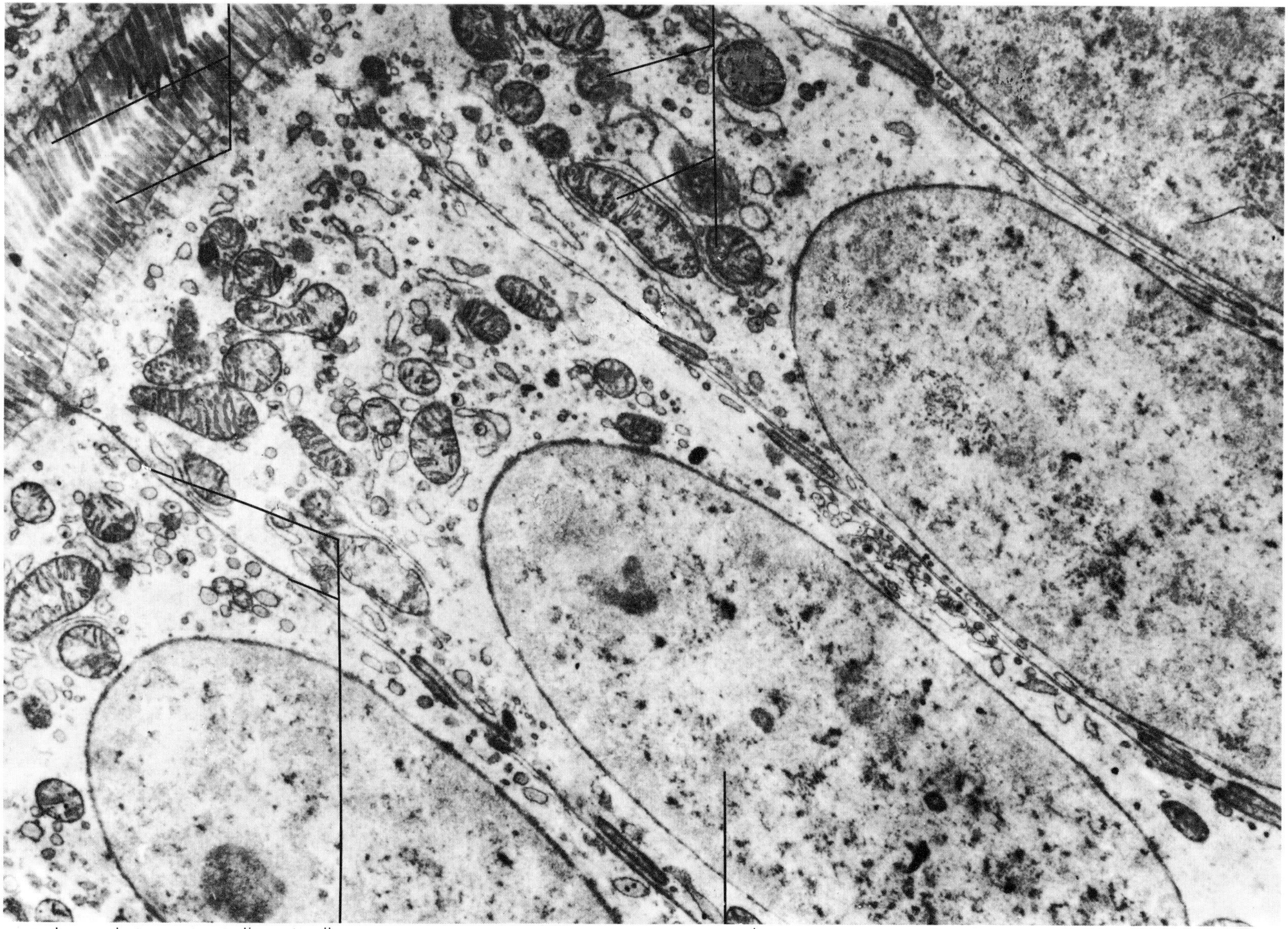
microvilli on outside of cells
mitochondria
membranes between two adjacent cells
nucleus

FIGURE 5·12

82

In the early nineteenth century nothing was known about the functions of the liver. The first scientific investigation of its function was made by the French physiologist Claude Bernard between 1849 and 1855. He became convinced that it was connected with the circulation of sugar in the body.

It had been supposed before this time that sugar was always taken in in the diet; but Bernard made the following observations:

I Glucose is always found in the blood and liver of carnivorous mammals (i.e. mammals which eat only meat).

II Blood entering the liver of carnivores by the hepatic portal vein contains no glucose, but blood leaving it by the hepatic vein always contains appreciable amounts.

What conclusions can you draw from these observations?

83

After Bernard had made these observations, it became commonly accepted that the liver produces sugar by acting upon something contained in the blood *at the very moment* that the blood passes through it. It was as if the blood were passing over a catalyst (the liver), which acted upon something it contained and caused the immediate appearance of sugar in it.

Bernard showed that this was not so. He took a liver from a freshly killed dog that had eaten tripe (the reticulum of a cow's stomach, i.e. a meal containing no sugar or starch) and washed out all its blood vessels by passing a stream of water for 40 minutes into the hepatic portal vein and out of the hepatic veins. At the start of the washing, the water coming out of the hepatic veins was sweet and produced a precipitate when heated with Fehling's solution. At the end, neither the liquid coming out of the hepatic veins nor a piece of the liver itself gave a precipitate when heated with Fehling's solution.

He then left the liver at room temperature for 24 hours, in a jar to prevent it from drying out. At the end of this time it contained 'sugar in abundance'. When water was squirted into the hepatic portal vein with a syringe, the liquid emerging from the hepatic vein gave a precipitate when boiled with Fehling's solution and was fermentable by yeast. Similar results were obtained from a part of the liver itself when it was boiled with Fehling's solution.

Clearly the fresh liver contained soluble sugar which could be washed out of it.

A What comment does this experiment make on the idea that the liver functions as a catalyst producing sugar from a substance contained in the blood as the blood flows over it?

B What does the experiment indicate about other materials connected with sugar that are contained in the liver?

84

Bernard made the following further observations:

I The formation of sugar in the washed liver stopped after 24 hours, though the liver cells were still alive at the end of this time. In other words, if the liver was re-washed after standing for 24 hours, no more sugar subsequently appeared in it.

II If a portion of washed liver was boiled immediately after washing, no sugar subsequently appeared in it.

What further information does this provide about the production of sugar by the liver tissue?

85

In another experiment, Bernard rubbed the washed liver through a fine strainer so that he obtained a pulp. This he dried gently at 40°C, so obtaining a dry liver powder. When water was mixed with the powder, sugar appeared in it after two hours. If the powder was boiled with water and then allowed to stand, no sugar appeared. We may assume that the sieving and drying at 40°C killed all the liver cells.

What further information does this experiment provide about the production of sugar by the liver tissue?

86

Bernard concluded that the liver does not produce sugar from a substance in the blood as the blood passes over it but that it constantly manufactures an insoluble substance which can subsequently be converted into sugar and released into the blood stream.

What is this insoluble substance? What kind of chemical is it?

87

The questions you have just answered give an accurate idea of the manner in which Bernard wrote up his experiments.

What in general is strikingly different about his method of investigation and presentation compared with present day methods (such, for example, as those quoted in the questions on oxygenation of the blood, 32, 33, FIGURES 2.7, 2.8, 2.9, 2.10, pp 35–37)?

88

Sugar, the source of energy for cell activity, is stored in plants mainly as starch. In animals it is stored in the liver and in muscles as glycogen, a starch-like substance.

What advantages can you think of to storing sugar in cells in the form of insoluble starchy material rather than as sugar itself?

89

The liver has many functions. We have examined one of them. Another is revealed by the following experimental facts. Read them carefully and see how much you can guess from them about this other function.

I An experimental dog has the blood supply to its liver short-circuited (so that no blood enters or leaves the liver), or has its liver removed. It survives and feeds for a few days but during that time no urea appears in its urine.

II The liver is removed from a dog and is fitted up with an artificial circulation. It is kept at body temperature and saline solution containing oxygen and glucose, having the same osmotic pressure as the blood, is passed through it. The liquid leaving the liver contains no urea.

When amino acids or a soluble source of ammonia are added to the saline entering the liver, urea appears in the liquid which leaves it.

90

The essential components of a balanced human diet are proteins, fats, carbohydrates, vitamins, mineral salts, water and roughage.

A **Proteins can be looked on as 'the stuff of life itself'. They are the structural materials from which cells are made.**

Also, all chemical reactions in the living cell are affected by certain groups of proteins. What is the name given to these proteins, and in what general way do they act?

B **There is a great number of proteins and their structure is very varied; even so there is a general test, Millon's test, which shows the presence of all proteins. Similarly, although there are quite a number of different reducing sugars, there is a general test which reveals their presence, Benedict's or Fehling's solution. There is no corresponding general test for vitamins. Why is this?**

91

In the early seventeenth century it was discovered by chance that the disease of scurvy, common among sailors who spent long periods at sea in sailing ships, could be prevented by eating citrus fruits. This knowledge did not produce any advance in man's knowledge of vitamins or of deficiency diseases, and vitamin C was not isolated from lemons until 1932.

Another early observation on deficiency diseases was made by the French investigator Magendie about 1800. He was the first to try to keep animals on a diet of purified foods, and it was the later perfection of this technique which did lead to advances in knowledge about deficiency diseases.

He fed dogs on a diet consisting of sugar, butter, gelatin (a protein) and distilled water. They all died in 30-36 days.

For which of the following reasons do you suppose Magendie included sugar, butter, gelatin and distilled water in the diet? Say also why you think the other two answers are not correct.

a These foods represent four of the main constituents of the diet.

b These foods are known to be essential components of living cells.

c These foods were known to contain no vitamins.

92

The first meaningful steps in obtaining an understanding of deficiency diseases came from the study of beri-beri. This disease was, and still is, very prevalent in Asia. Some of the symptoms of it are disorder of the nerves controlling the limbs, followed by wasting of the limbs, staggering and inability to make any purposeful movement of the hands or arms. The heartbeat is disturbed, and may fail altogether.

The Japanese physician Takaki showed in 1887 that this disease did not appear in naval crews if they were given barley in their diet instead of the customary polished rice. Half the world's population did and does depend on rice for its survival. 'Polished' rice is the sort used for making rice puddings, that is the outer husk, which is not pleasant to eat, has been removed.

In the 1890s, Pasteur's work on bacteria and bacterial diseases dominated medical thought. The Dutch government sent a commission to the Dutch East Indies to investigate beri-beri, and they spent two years searching in vain for a micro-organism which might cause it.

In 1897, Eijkman, a member of this commission, observed that experimental chickens fed on food scraps of polished rice soon developed beri-beri. He found by experiment that if they were given unpolished rice, or if the rice husks were included with the polished rice, they did not develop beri-beri and concluded that the husk 'contained a substance or substances which neutralized the harmful influence of the starchy food'.

Obviously no civilized person can experiment on human beings; but Eijkman saw an unusual opportunity for gathering data. In Java and the neighbouring island of Madura, there were 100 prisons. In some of these it was the custom to feed the prisoners with polished rice, in others with unpolished rice. Eijkman enquired about the incidence of beri-beri in these prisons. The results appear in the table below (FIGURE 5.13). Since it was possible that the disease was related to prison conditions other than the diet, Eijkman enquired also the age of the prisons, their ventilation and their degree of crowding. The figures he got are as follows:

Age of building	
40–100 years	beri-beri occurred in **13** out of **26** prisons, i.e. in **50**%
21–40 years	beri-beri occurred in **11** out of **32** prisons, i.e. in **34**%
2–20 years	beri-beri occurred in **19** out of **42** prisons, i.e. in **45**%
Ventilation	
Good	beri-beri occurred in **28** out of **68** prisons, i.e. in **41**%
Medium	beri-beri occurred in **8** out of **11** prisons, i.e. in **73**%
Bad	beri-beri occurred in **7** out of **21** prisons, i.e. in **33**%
Crowding of prison	
Uncrowded	beri-beri occurred in **32** out of **73** prisons, i.e. in **45**%
Medium	beri-beri occurred in **1** out of **1** prisons, i.e. in **100**%
Overcrowded	beri-beri occurred in **9** out of **26** prisons, i.e. in **35**%
Diet	
Rice partly polished, but at least part of the husk left on.	
	beri-beri occurred in **1** out of **35** prisons, i.e. in **2.7**% (proportion of prisoners with beri-beri 1 in 10 000)
Mixture of polished and unpolished rice	
	beri-beri occurred in **6** out of **13** prisons, i.e. in **46**% (proportion of prisoners with beri-beri 1 in 416)
Polished rice	beri-beri occurred in **36** out of **51** prisons, i.e. in **71**% (proportion of prisoners with beri-beri 1 in 39)

FIGURE 5·13

What deductions can you make from these data?

93

The anti beri-beri vitamin is called vitamin B1, or thiamine. Relatively large amounts of it are found in the heart. This may indicate that it plays an important part in the metabolism of the heart. The rate of heartbeat in rats slows down when they are fed on a diet lacking vitamin B1 (FIGURE 5.14).

A i For how long after vitamin B1 is withdrawn from the diet does the heartbeat remain normal?

ii In view of the fact that the rate of heartbeat rapidly declines after this, why do you think it remains steady for the period you have just stated?

B What is the effect of adding thiamine to the diet at the point indicated (about 33 days after the start of the experiment)?

C Suppose we had 10 rats in the condition shown in the graph after 30 days of the experiment (i.e. with average rate of heartbeat of just over 250 per minute). These animals could be used to detect the amount of vitamin B1 in different foods by feeding them with known weights of food and measuring the increase in rate of heartbeat.

Which of the four following experimental difficulties in this experiment do you think is likely to be the EASIEST **to control?**

a **Variation in the appetite of the rats, affecting the amount of vitamin B1 eaten.**

b **Variation between rats in the response of the heartbeat to the same doses of vitamin B1.**

c **Variation between rats in the amount of vitamin B1 absorbed into the blood from the food.**

d **Variations in the mechanical apparatus used to measure the rate of heartbeat.**

94

If the rate of heartbeat of a rat used in this experiment were 375 per minute, what would you expect to be the concentration of vitamin B1 in the rat's diet, expressed as a percentage of the concentration which would produce normal rate of heartbeat?

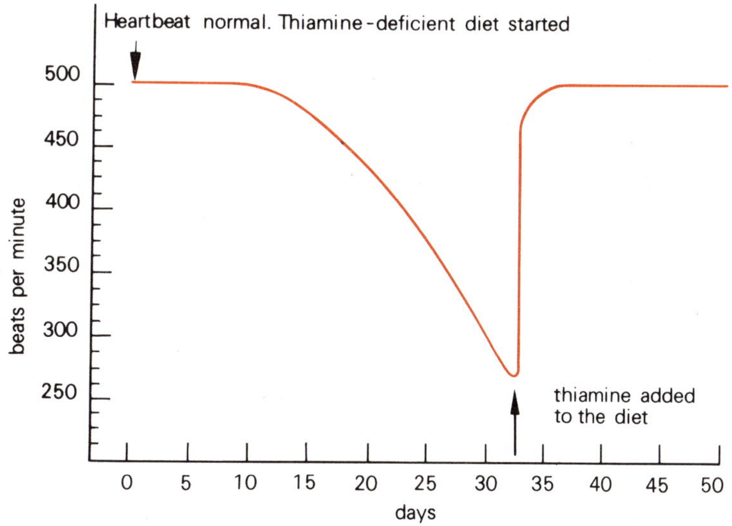

FIGURE 5·14

95

FIGURE 5.15 shows the results of a controlled-feeding experiment like Magendie's in which more accurate measurements were made.

The English biochemist Hopkins kept two groups of eight young male rats on a diet consisting of pure protein, starch, sucrose, inorganic salts and water. Group A were given in addition a small daily ration of milk, representing about 4% of the total food eaten. Group B were given exactly the same diet except that they received no milk. After eighteen days the milk was switched from group A and given to group B.

A **Why was it important to carry out this experiment on groups of rats rather than on two individual rats?**

B **What was the average weight of the rats at the start of the experiment?**

C **Consider group** A alone **and group** B alone. **What was the effect of milk and the absence of milk on the growth of the rats in group** A **and group** B **as revealed by this graph?**

D **Consider the first 18 days of the experiment. Hopkins had two groups of rats,** A **and** B. **Because of this he could draw a conclusion from their growth rates that he could not have drawn had he only had group** A. **What is this conclusion?**

E **Group** B **increased in weight for a while after the start of the experiment. Why do you think this was so?**

F **Cow's milk contains at least ten vitamins and is an important source for human beings of two of them. Which are these two?**

G **Which one of the following conclusions are we best justified in drawing from the information contained in the graph? Explain why the other conclusions are not as good.**

a **That milk in the diet will promote the growth of rats and that absence of milk will retard it.**

b **That milk contains a vitamin essential to the growth of rats.**

c **When rats are kept on the diet with no milk for 18 days and are then given milk, there is an immediate increase in the average weight.**

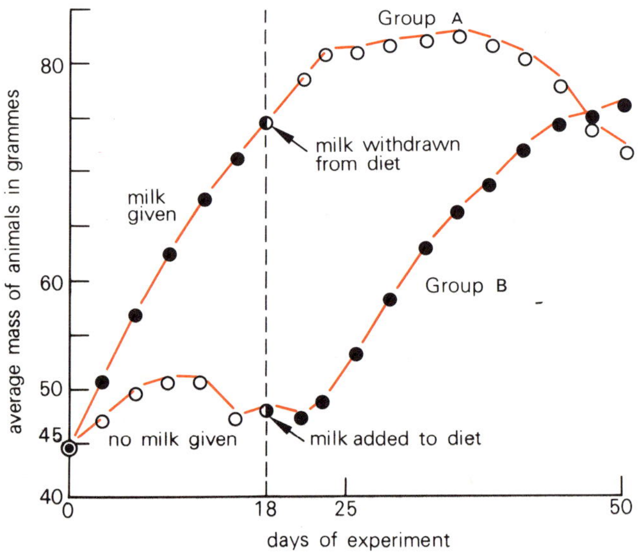

FIGURE 5·15

96

With what chemical process in cells are the B group vitamins thiamine, riboflavine and nicotinic acid concerned?

97

There are about twenty different amino acids; not all proteins contain all of them. On average there are some 500 amino acid units in a protein. The number of units of any particular amino acid and the order in which the different units are arranged varies from protein to protein.

There is an almost infinite number of possible arrangements of amino acid units in different proteins, and probably every species of animal and plant contains slightly different proteins from other species. When a man eats cow, pig or sheep protein he digests it. It is insoluble, and so could not be carried in the blood stream to the growing tissues. Also, the animal protein is not quite the right sort for making into human tissue. In digestion it is dismantled into its constituent amino acids, and these are reassembled in human cells (under instructions from the DNA in the nucleus) into human proteins.

Some of the amino acids can be made in the human body from ammonia and carbon compounds. Others (about ten of the twenty different kinds) either cannot be made at all or at least not fast enough; they must therefore be eaten if the person is to survive.

FIGURE 5.16 shows the results of a nutritional experiment with *young* rats. Those in group I were fed on a diet containing all necessary vitamins, carbohydrates, fats, mineral salts, water and roughage; but the only protein in it was *zein*, a protein found in maize (sweet corn). A second group (group II) were fed on the same diet but after seven days, two amino acids, lysine and tryptophan, were added in small quantities as well.

A **What was the effect of feeding young rats on a diet containing only zein as a source of protein?**

B **What was the effect on growth of adding lysine and tryptophan to the diet?**

C **i** **What conclusion might we draw about the composition of the protein zein?**

ii **What conclusion might we draw about the ability of young rats to manufacture lysine and tryptophan?**

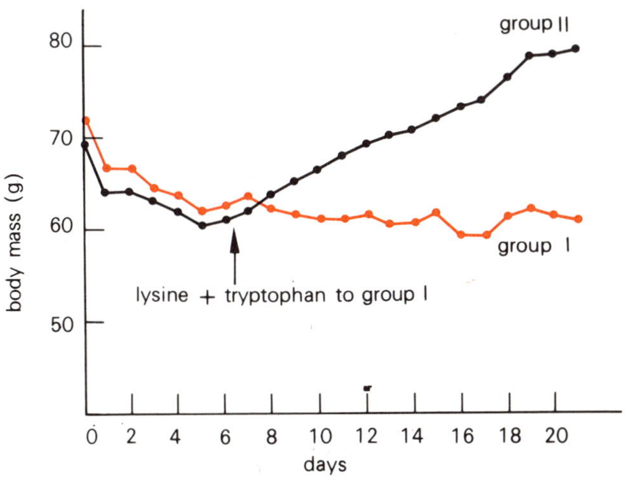

FIGURE 5·16

98

Those amino acids which the body either cannot make at all or cannot make fast enough to satisfy the demands of the tissues are called *essential* amino acids. Ten amino acids are known to be essential for human beings.

FIGURE 5.17 shows the effects of withholding one amino acid, valine, from the otherwise complete diet of a rat. (Unfortunately the original photograph from this classic experiment has been lost; the drawings here are copied from it.) The upper drawing shows the animal after 28 days on this diet; it had lost one third of its body weight and was near death. It was then given valine in its food. The lower drawing shows it 25 days later.

The above definition of an *essential* amino acid is strictly correct, but one must not think in too black and white a manner about 'essential' and 'non-essential' amino acids. Take the following example:

Methionine is an essential amino acid for rats. They cannot manufacture it. If it is not eaten in the rat's diet, deficiency symptoms, like those in FIGURE 5.17 develop.

Cystine is a non-essential amino acid; the tissues can manufacture cystine from methionine and this is one of the uses made of methionine.

FIGURE 5·17

If a rat is given a diet complete except that it contains slightly too little methionine, it will develop deficiency symptoms. If the amount of methionine is unchanged but more cystine is added to the diet, the deficiency symptoms do not appear.

A **Why do you think that the non-essential amino acid cystine is able to prevent the appearance of methionine deficiency symptoms?**

B **Taking the definition of an essential amino acid given at the beginning of this question, is cystine to be regarded as an essential amino acid or not? Give the reason for your answer.**

99

FIGURE 5.18 shows the effect of withholding lysine from the diet of *adult* rats. (Note the weights of the rats in this experiment compared with those in FIGURE 5.16.) Rats in group A were fed on a complete diet in which the protein source was zein plus lysine plus tryptophan. Rats in group B had the same diet except that lysine was omitted.

A **What does this graph tell you about the ability of adult rats to synthesize lysine?**

B **What does this graph tell you about the ability of adult rats to synthesize tryptophan?**

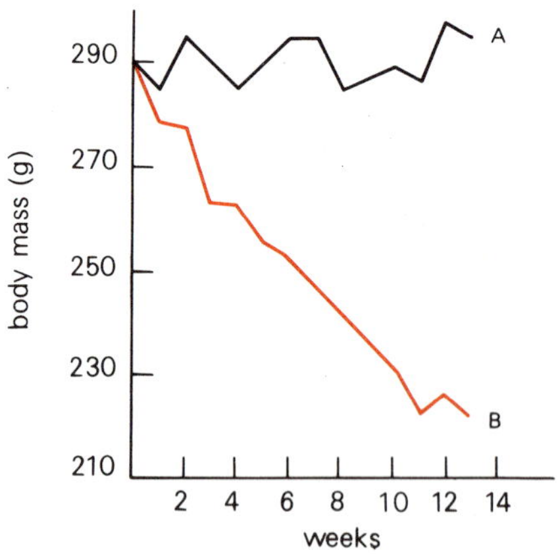

FIGURE 5·18

100

Protein is used in the body for building new cell material. An adult European takes into his bloodstream from the gut about 100g of amino acid per day. Of this about 20g are used for replacement of dead cells. The remaining 180g is *deaminated* in the liver. This means that it is converted into carbon compounds which can be used as energy sources, and urea (chemical formula $CO(NH_2)_2$). All the nitrogen of the amino acid ends up in urea, which is excreted.

Thus most of the protein which is eaten is used not as building material but as an energy source, as are fats and carbohydrates.

1 g of protein when oxidized in the body yields 4 kilocalories.

1 g of fat when oxidized in the body yields 9 kilocalories.

1 g of carbohydrate when oxidized in the body yields 4 kilocalories.

FIGURE 5.19 shows the daily diet of women doing light engineering work in Birmingham in 1943. COLUMN A shows the kilocalorie content of each meal; COLUMNS B and C show the protein and fat content of each meal.

MEAL	A kcal	B PROTEIN (g)	C FAT (g)	D kcal FROM PROTEIN	E kcal FROM FAT	F kcal FROM CH_2O	G gCH_2O IN MEAL
Breakfast, 7 a.m.	710	21	26				
Snack, 10 a.m.	210	9	8				
Dinner, 1 p.m.	830	30	27				
Snack, 4 p.m.	240	5	9				
Tea, 7 p.m.	570	19	21				
Total day's nutrients	2560	84	91				

FIGURE 5·19

A **Assuming for simplicity that the protein eaten is used entirely as a source of energy, copy the table and fill in** COLUMN D, **showing how many kilocalories were derived from the protein content of each meal.**

B **Do the same for fat in** COLUMN E.

C **Calculate the number of kilocalories that must, therefore, have come from carbohydrate eaten at each meal; enter the values in** COLUMN F.

D Calculate the number of grammes of carbohydrate eaten at each meal and enter the values in COLUMN G.

101

FIGURE 5.20 is an analysis of the major constituents of two meals. The first is a snack 'packed lunch' type, which many office workers take regularly. The second appears to be a more substantial 'good hot meal', of the kind served in factory canteens and school dining halls.

MEAL 1	WEIGHT (g)	kcal	PROTEIN (g)	FAT (g)	CALCIUM (mg)	IRON (mg)
Roll	98	252	8·8	1·1	109	1·8
Butter	8	65	0·0	7·3	0	0·0
Cheese	56	234	14·2	19·6	460	0·4
Tea (with milk and sugar)	280	20	1·0	1·0	30	0·0
Total	442	571	24·0	29·0	599	2·2

MEAL 2	WEIGHT (g)	kcal	PROTEIN (g)	FAT (g)	CALCIUM (mg)	IRON (mg)
Mutton	67	225	8·9	21·1	7	1·4
Cabbage	84	21	1·2	0·0	54	0·9
Potato	112	84	2·4	0·0	8	0·8
Apples	84	36	0·3	0·0	3	0·3
Custard	56	62	1·8	2·2	70	0·0
Total	403	428	14·6	23·3	142	3·4

FIGURE 5·20

Look at each column in turn and say for each item which of the two meals is superior.

102

Until the late eighteenth century very little was known about the ways in which the activities of the human body are coordinated. The Greek physician Galen had suggested about the year A.D. 200 that the brain created 'animal spirits' which were conveyed along the nerves to the muscles, causing them to contract. This view was held until nearly 1800, which may seem surprising, but the art of thinking in scientific terms did not occur to man till about this time, and in particular the electric cell was not invented until 1800 (by Volta). No serious study of electrical phenomena could start before this, which ruled out any advance in knowledge of how nerves work.

Thereafter advance was rapid, and by the mid-nineteenth century a correct picture of the way in which the nervous impulse is transmitted had been obtained.

In 1849 a German, Berthold, performed an experiment with cockerels which gave the first clue that as well as nerves, the body produces chemical coordinators. The biological world was not ready for this idea, and the study of hormones and endocrine glands began in earnest seventy years later, in about 1920.

In this chapter we are going first to learn about some of the early investigations on hormones. Then, when we have some picture of their activity, we shall see what generalizations we can make about them.

It is common practice to *caponise* young cockerels, that is to castrate them, or remove their testes. This makes them less aggressive, and their muscles do not grow so tough as those of cockerels.

The possession of either testes or ovaries is the primary sexual characteristic, and determines whether an individual is male or female. There are as well secondary sexual characteristics. In cockerels (FIGURE 6.1a) these are large heads bearing fleshy growths called combs and wattles, an aggressive fighting attitude to other cockerels, a display behaviour towards hens, and the cock-a-doodle-do cry.

Castrated cockerels are timid, do not have combs or wattles, do not fight or display to the female, and do not crow. They are called *capons* (FIGURE 6.1b).

FIGURE 6·1a

FIGURE 6·1b

Berthold used four young cockerels. They were treated as follows:

Bird I Both testes were removed. The animal developed into a capon. When killed four months later, it was found to contain no testes.

Bird II Both testes were removed. One was replaced among the intestines. Of course, removing it from its original position had cut off its blood and nerve supply. The bird developed into a cockerel and when killed and examined two months later, the testis was found attached to the wall of the intestine. It had grown, and contained sperms.

Bird III This was castrated and one testis was replaced as before.

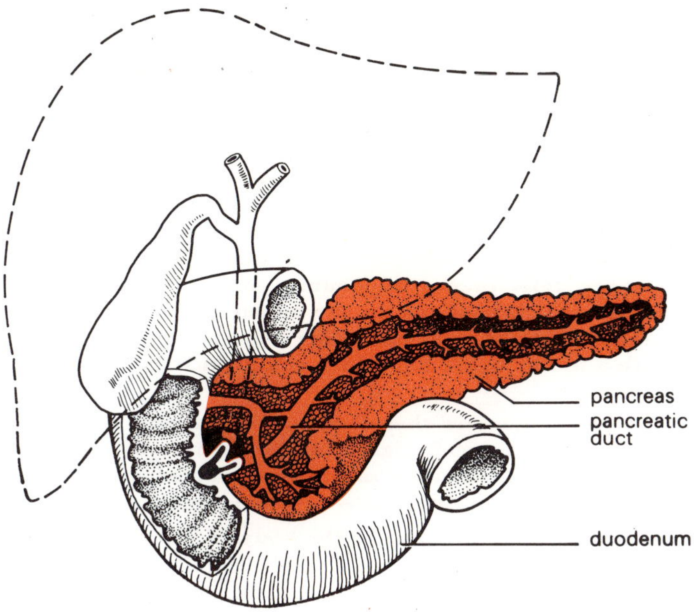

FIGURE 6·2

It too developed into a cockerel. After two months, the comb and wattle which it had grown were cut off. The testis was found growing attached to the wall of the intestine. It was removed. The bird thereafter behaved as a capon, did not fight other cockerels or display before hens, did not crow and did not grow the comb and wattle again.

Bird IV As before, the animal was castrated and one testis was replaced among the intestines. Two months later the comb and wattle were removed but the testis was left in place on the wall of the intestine. The comb and wattle grew again, and the bird continued to behave as a cockerel.

A **The fact that** *bird I* **had no testis four months after the operation shows that the testis cannot be regenerated by the body. Neither can a severed nerve supply be regenerated. Yet the testis in** *bird II* **grew and produced sperm.**

What does this indicate about its blood supply?

B *Bird I*, **with no testes, developed into a capon;** *bird II*, **with a testis transplanted on to the outer wall of its intestine, grew into a cockerel. What does this tell us about:**

i the relationship between the testis and the secondary sexual characteristics?

ii the location of the testis with regard to the development of the secondary sexual characteristics?

C **What connection did the transplanted testis in** *bird II* **have with the rest of the body that could allow it to influence sexual development?**

D **Suppose the testis produces a chemical substance which affects sexual development. What does the experiment in transplantation in** *bird II* **tell us about the influence of nerves on the production of this substance?**

E **Comparison of the results on** *birds I* **and** *II* **shows the effect of the testis on the development of secondary sexual characteristics in the young, growing animal. What does the effect of removing the testis from the** *adult cockerel III* **reveal about the effect the testis has on the** retention **of secondary sexual characteristics by the adult animal?**

F **What does the behaviour of** *bird IV* **reveal about the activity of the testis in the adult animal?**

103

Berthold's work did not lead to rapid developments in the realm of hormones and chemical control of body processes. Biologists were too busy concentrating on the study of nervous control and working out reflex actions, the organization of the spinal cord and the brain, to pay attention to the idea of chemical coordination.

The next big advance came in 1902 from experiments by Bayliss and Starling on the control of the secretion of the pancreas.

The pancreas lies in the loop of the duodenum (FIGURE 6.2), into which it opens by the pancreatic duct. It has a rich blood supply and receives nerves from various sources.

It contains two types of cell, which can be seen in FIGURE 6.3.

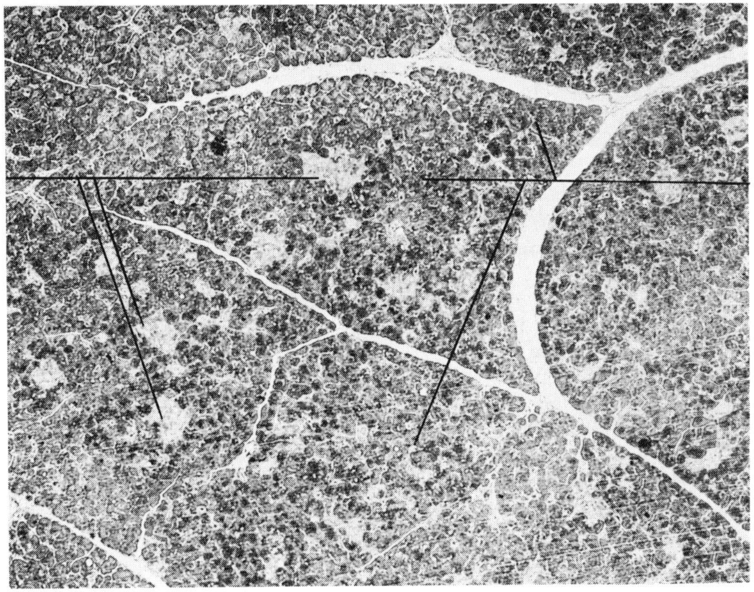

FIGURE 6·3

Most are of the darkly-staining type, and these produce *pancreatic juice*, a liquid rich in digestive enzymes which act on proteins, carbohydrates and fats. Between meals there is a continuous but small secretion of pancreatic juice. Within one or two minutes of starting a meal, a free flow of juice starts which continues for about three hours.

The Russian biologist Pavlov, who contributed a lot to man's knowledge of nervous reflexes, thought that secretion of pancreatic juice was a reflex action. He suggested that the wall of the duodenum contained cells sensitive to the presence of food: when these were stimulated, a message was taken into the spinal cord by a sensory nerve, resulting in a message passing out of the cord to the pancreas, causing it to secrete pancreatic juice.

It soon became known that hydrochloric acid introduced into the cavity of the duodenum or small intestine caused a rich flow of pancreatic juice about two minutes later, and it seemed that the strongest stimulant in promoting pancreatic flow was the acid stomach contents passing the pyloric sphincter into the duodenum.

A **In their first experiment, Bayliss and Starling, experimenting with a dog, cut off all the nerve supply from the brain and spinal cord to the pancreas and duodenum. On injecting 30 cm^3 of 0·4% hydrochloric acid into the cavity of the duodenum, a marked flow of pancreatic juice followed after an interval of two minutes.**

What does this experiment tell us about the control of the secretion of pancreatic juice?

B **The pancreas contains several nerve ganglia, that is, small swellings which contain nerve cell bodies. It is possible that the flow of pancreatic juice is controlled not by a reflex arc which operates via the brain or spinal cord but by one operating outside it. For example, there could be sensory cells in the duodenum connected with sensory nerves which pass not to the spinal cord but direct to a ganglion on the pancreas. There the nerve could synapse with a motor nerve cell which could cause pancreatic cells to secrete.**

The experimenters therefore isolated a loop of the small intestine by tying it tightly at both ends. All the nerves serving this loop were carefully dissected away, but the blood vessels passing to it were undamaged.

The loop was connected to the body of the animal only by arteries and veins.

On injecting 19 cm^3 of 0·4% hydrochloric acid into the loop, a rich flow of pancreatic juice followed two minutes later.

i **What was the only connection between the pancreas and the small intestine in this experiment?**

ii **Suggest two possibilities for the way in which hydrochloric acid was able to produce a flow of pancreatic juice.**

iii **It was known that the introduction of hydrochloric acid into the blood stream did not have any effect on pancreatic secretion. How does this modify your answer to ii?**

iv **Why did the experimenters tie off the loop of the intestine which was being used?**

C **They next cut out the loop of the small intestine, scraped off the cells lining its surface, rubbed them with 0·4% hydrochloric acid in a mortar, filtered the extract and injected it into one of the dog's veins. After 70 seconds there was a copious flow of pancreatic juice.**

i **What does this experiment reveal about the origin of a**

chemical substance which can be carried by the blood to the pancreas, causing it to secrete pancreatic juice?

ii Suggest a control for this experiment.

iii The experiment described in C was regarded as conclusive proof that reflex nervous control was not involved in the secretion of pancreatic juice. Explain why this is so.

104

Later, in 1927, another scientist, Ivy, added more weight to the anti-reflex control argument.

He operated on a dog and moved a portion of its pancreas and a portion of its duodenum to two positions beneath the skin. Both pieces of tissue become grafted into the new positions. Blood vessels from the skin grew into them. Then the nerves and blood supply connecting the pancreatic piece to the pancreas and the duodenal piece to the duodenum were cut. The two pieces then had no nerve supply at all.

When hydrochloric acid was introduced to the duodenal piece, the pancreatic piece secreted pancreatic juice.

Explain why this experiment was an improvement of that of Bayliss and Starling, and why its results completely rule out the possibility of secretion being produced by a nervous reflex.

105

In 1889, two biologists, Mering and Minkowski, carried out an experiment which was to benefit millions of human beings. They operated on a dog and removed its pancreas. The animal recovered and lived for a time. It appeared to be able to manage without the enzymes of the pancreatic juice (presumable the disgestive juices of the stomach and intestine made up for them), but it died a month or so later of a disease called *diabetes mellitus*.

Diabetes mellitus is a disease as ancient as man's history. Its symptoms are that the amount of glucose in the blood rises from the normal level of 0·1% to about 0·2–0·5%, and glucose appears in the urine (it is very seldom present in the urine of normal persons). Since glucose is being lost, the glycogen reserves are rapidly depleted. Sugar is made from muscle protein and from body fat, and the patient starts to waste away. The rapid breakdown of large amounts of fat results in the formation of poisonous chemicals, which are probably the immediate cause of death. In the U.S.A. alone there are half a million diabetics. Before 1920, a diabetic could only expect to live a year or two. Today his expectancy is little short of that of a normal person.

Before 1889, the cause of this disease was a mystery; but Mering and Minkowski found that all the symptoms listed above developed in the dog whose pancreas they had removed. Clearly the pancreas not only produces digestive juices but also in some way keeps the level of blood sugar at the normal 0·1% and prevents all the secondary symptoms of diabetes.

Two questions had to be investigated.

a Which part of the pancreas has this effect?

b How does it have this effect?

An obvious first guess was that the pale patches of cells which can be seen in FIGURE 6.3, called the Islets of Langerhans, might be responsible. There are between 200 000 and 2 000 000 of these patches in an adult human, and they make up about 1% of the total weight of the pancreas.

A In the next thirty years many biologists tried to investigate this possibility. It was discovered that if a thread was drawn round the pancreatic duct (see FIGURE 6.2), tying off and blocking it, the dark-staining cells (FIGURE 6.3) all died but the pale 'island' cells were unaffected. An animal treated in this way remained healthy and did not develop diabetes.

What does this indicate about the function of the dark-staining cells and of the island cells in affecting blood sugar level?

B In another experiment with a dog, the ducts leading from the bottom third of the pancreas were tied off but those from the rest of the organ were left untouched. The animal showed no ill effects. After seven months, the untouched two thirds of the pancreas were removed. Still the animal showed no ill effects, in particular no diabetes. Three weeks later, the final third of the pancreas was removed. The dog rapidly developed diabetes and died.

What extra support does this experiment give to the suggestion that the island cells are responsible for maintaining the normal blood sugar level in the animal?

106

Thus it was established that the island tissue of the pancreas was the part responsible for controlling blood sugar level. The next question was how it did it.

Two suggestions were made:

a that the blood was changed in some way while passing over the island tissue. This did not seem a very likely hypothesis since there was so little of the island tissue.

b that the island tissue produced a hormone which affected blood sugar level.

If it did produce a hormone, one might expect to be able to extract it from the pancreas and by injecting it into a diabetic animal, relieve the diabetes. This would be a similar procedure to that of Bayliss and Starling when they scraped off intestinal lining, mashed it with hydrochloric acid, injected the filtered extract into a vein and produced a flow of pancreatic juice.

Many biologists tried to do this by grinding the pancreas with sand and filtering the extract. No one succeeded until in 1920 Banting and Best hit on the idea that since the dark-staining tissue of the pancreas produces digestive enzymes, these, when mashed together with hormone-producing cells, might be destroying the hormone.

They therefore decided that instead of trying to extract the hormone from an intact pancreas, they would first tie off the pancreatic duct, allow the dark-staining tissue to die, and then try to extract hormone by mashing island cells and filtering the extract. The island cells would be the only ones remaining in the pancreas. This they did. The results have been described as 'one of the most important events in medical history'.

FIGURE 6.4 is a reproduction of the original results which Banting and Best obtained in 1922. The Y axis shows blood sugar level. The normal level for the dog was 0·09%, i.e. 90 mg glucose per 100 cm³ blood. The X axis shows time in two hour intervals.

A The dog's pancreas was removed at 3 p.m. on 3 August. What happened to the animal's blood sugar level during the next 17 hours?

B What happened to the blood sugar level over the subsequent five hours?

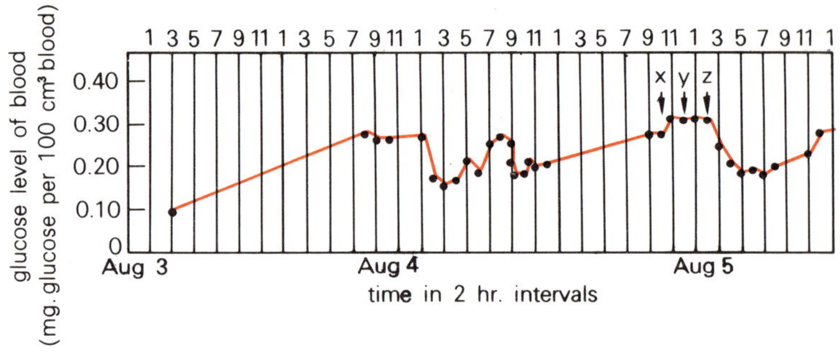

FIGURE 6·4

C At 1 p.m. on 4 August (22 hours after the operation) the dog was given an injection of 5 cm³ pancreatic extract prepared as described above (i.e. prepared from the island cells only). What effect did it have on the animal's blood sugar level in the following two hours?

D Describe the course of blood sugar level over the next 19 hours. (A second injection of 5 cm³ pancreatic extract was given at 9 p.m. on 4 August.)

E i How did the effect of the pancreatic extract on blood sugar level compare with that of the intact, living pancreas?

ii How do the results help us to distinguish between the two hypotheses advanced above for the possible action of the island cells in controlling blood sugar level? (The hypotheses were:

a that the blood was changed in some way when passing over the island cells

b that the island cells produced a hormone which affected blood sugar level.)

iii In this experiment, pancreatic island extract was injected into the bloodstream and was distributed to all parts of the dog's body. What, then, would you expect to be the mode of action of the island cells in the living animal in maintaining normal blood sugar level?

F If the hormone, which Banting and Best called insulin, **was delivered to all parts of the body, one might expect to find it in, and perhaps be able to extract it from, other organs, provided it had not been destroyed by enzyme or other action.**

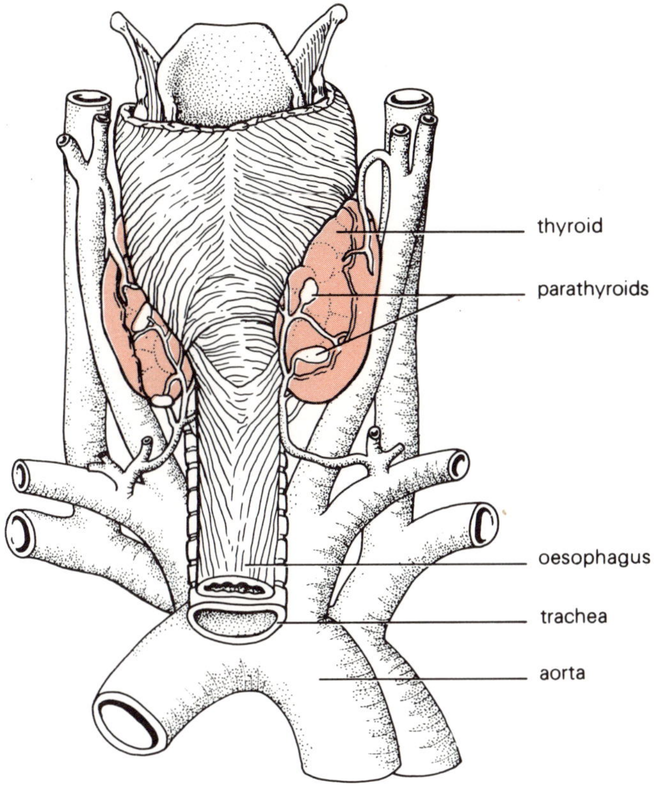

FIGURE 6·5

Therefore at the points X and Y in FIGURE 6.4, the experimenters injected 5 cm³ of a liver b spleen extract. At the point Z they injected a further 5 cm³ of pancreatic extract.

What effects did these injections have, and what do they tell you?

G The experimenters now mixed 20 cm³ of pancreatic extract with 10 cm³ of active pancreatic juice obtained from a normal pancreas which had not been operated on. The mixture was kept at body temperature for three hours and then injected into the experimental dog. No drop in blood sugar resulted. Suggest why this was so.

107

Much has been learned about insulin since 1922, and much remains to be learned. Because it helps to control the way that cells use sugar, it is of basic importance to the life of the cell. Its effects are:

I Glucose circulates in the blood and passes into the living cells from the lymph. Insulin makes the passage of glucose into the cells easier, that is, it helps to regulate their supply of glucose.

II Insulin must be present for ATP, the universal energy donor, to be formed by the oxidation of glucose in the cell.

III Insulin causes the liver to produce glycogen from glucose.

A How do these effects of insulin fit in with the symptoms of diabetes described in question 105 ?

B Effects I and II promote the rate at which glucose is used by the cells of the body. This would be necessary, for example, under conditions of muscular exertion, when more energy would be needed. But for this effect to work, more glucose must be available in the blood at such a time.

What hormone do you know of that increases the amount of glucose available in the blood?

108

Another of the endocrine glands is the thyroid (FIGURE 6.5). It is shaped like a butterfly; in this figure we cannot see the portion of it which lies behind the trachea. In a healthy adult it weighs about 25 g.

Inside it consists of an enormous number of spherical compartments, each lined with cubically shaped cells and filled with semi-liquid material. This contains the hormones thyroxine and tri-iodo-thyronine. The thyroid is the only endocrine gland which has a store of its hormone. All the others produce hormones only as required, which means that only a very little can be extracted from them.

About 60% of the weight of the thyroxine and of the tri-iodo-thyronine molecule is iodine. The thyroid gland has a very good blood supply. About 150 cm³ of blood passes through it each minute. Thus any iodine which is taken into the body in the food (for example as iodized table salt) is rapidly exposed to the cells of the thyroid gland.

A FIGURE 6.6 **shows the effect of injecting 10 units of a radioactive iodine compound into rats. The proportion of radioactive iodine in the plasma (see Y axis on the right of the graph) and the proportion of radioactive iodine in the thyroid gland (see Y axis on the left of the graph) are shown in lines A and B.**

The 10 units of radioactive iodine were injected at time 0. Readings of the amount of iodine in the plasma and in the thyroid were taken after 2, 5, 12, 25 and 50 hours (see X axis).

i **At time 0 there were 10 units of radioactive iodine in the plasma and none in the thyroid gland. What was the situation five hours later?**

ii **What does this reveal about the activity of the thyroid gland?**

B FIGURE 6.7 **shows similar results from human patients. Three persons, one with a normally active thyroid gland, one with an over-active thyroid gland and a third with an under-active thyroid gland were given the same amount of radioactive iodine in their diet. The Y axis shows the percentage of radioactive iodine taken in which found its way into their thyroid glands.**

Describe the results of this experiment and say how they compare with those shown in FIGURE 6.6.

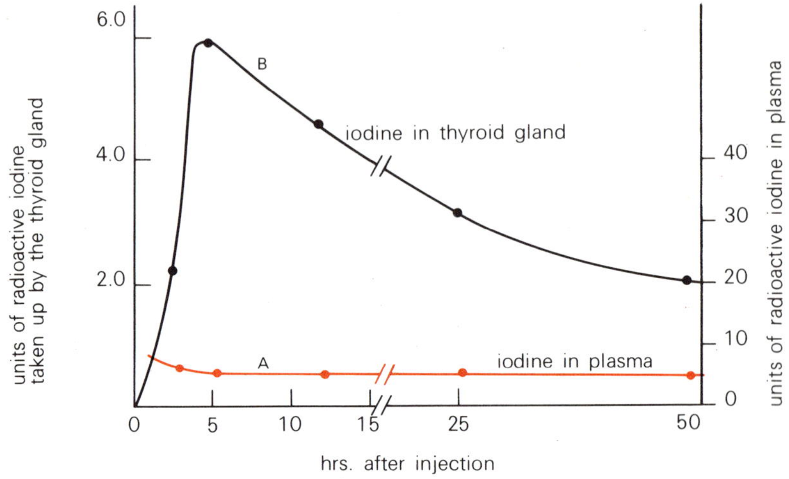

FIGURE 6·6

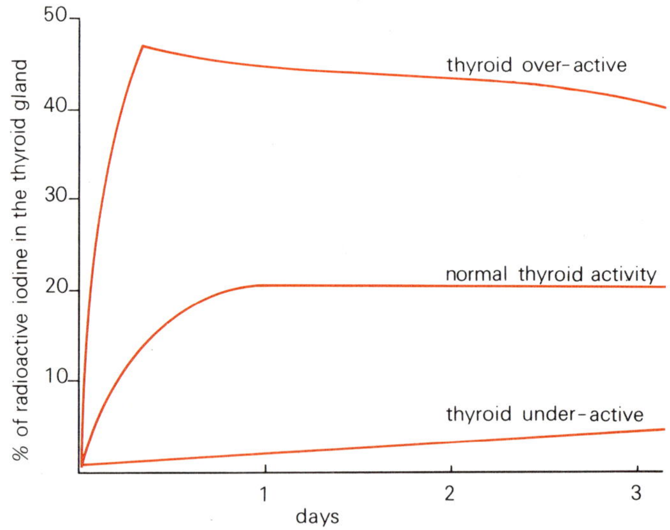

FIGURE 6·7

109

Like insulin, thyroid hormones have many effects on the body. One fundamental effect, like that of insulin, is on the oxidation of glucose, i.e. on the energy-liberating processes of cells.

The basal metabolic rate is the amount of energy expended when a person is in a state of complete mental and physical rest. It represents the energy needed to maintain body temperature at 37°C and to provide for such essential processes as heartbeat and breathing.

FIGURE 6.8 shows (period A) the basal metabolic rate of an adult human patient whose thyroid gland was not working.

A **How much less than normal was this person's basal metabolic rate?**

B **From the above definition of basal metabolic rate, what symptoms do you think this patient displayed?**

C **What was the effect on basal metabolic rate of the injection of thyroxine at the beginning of period** B **?**

D **What was the effect on basal metabolic rate of the injection of tri-iodo-thyronine at the beginning of period** C **?**

E **What part do you think that thyroxine plays in maintaining the body temperature of mammals? (Man, whose deep internal temperature is 37°C, lives in an environment in which the average temperature is 20–25°C.)**

F **The way in which thyroid hormones act on cells is poorly understood, though one of their effects is on the oxidation of glucose. This oxidation is coupled with the generation of ATP: some of the energy liberated is 'captured' as ATP and can be used for other chemical processes in the cell. A maximum of 45% of the energy from the oxidation can end up as ATP, the other 55% appears as heat.**

We can represent this simply as:

glucose ⟶ $6CO_2 + 6H_2O$ coupled with 36 ADP + 36 P ⟶ 36 ATP.

It seems that thyroid hormones 'uncouple' these reactions so that glucose is oxidized without **the formation of ATP.**

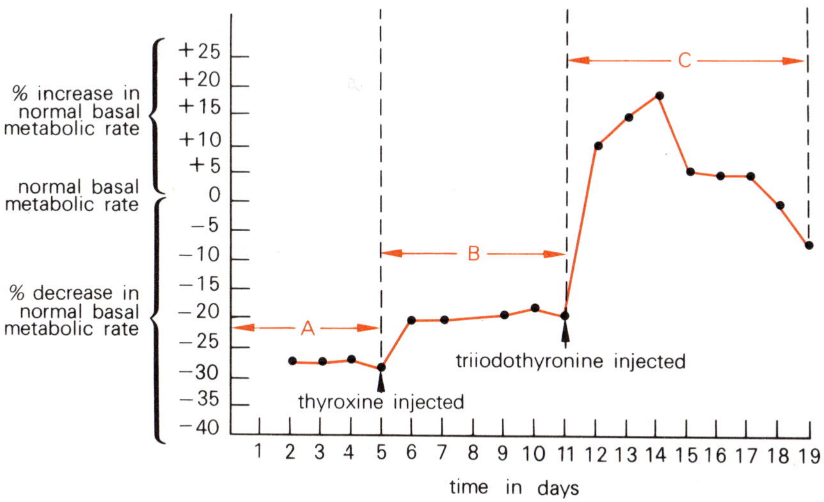

FIGURE 6·8

How would this help to explain the effect that thyroid hormones have on the basal metabolic rate?

110

The activity of the thyroid gland in producing thyroid hormones is regulated by another hormone, produced by the pituitary gland. It is called Thyroid Stimulating Hormone (TSH). TSH causes the cells in the thyroid gland to multiply, to produce more thyroid hormones and to release them into the blood.

Also, the output of TSH by the pituitary is governed by the level of thyroid hormones in the blood. When the level in the blood falls, the pituitary produces more TSH, this automatically causing an increased output of thyroid hormones. Thus secretion of thyroid hormones is self-regulating. If more thyroid hormone is used by the cells, the level in the blood falls, the pituitary secretes more TSH, and the thyroid secretes more thyroid hormones.

We can represent this as a diagram:

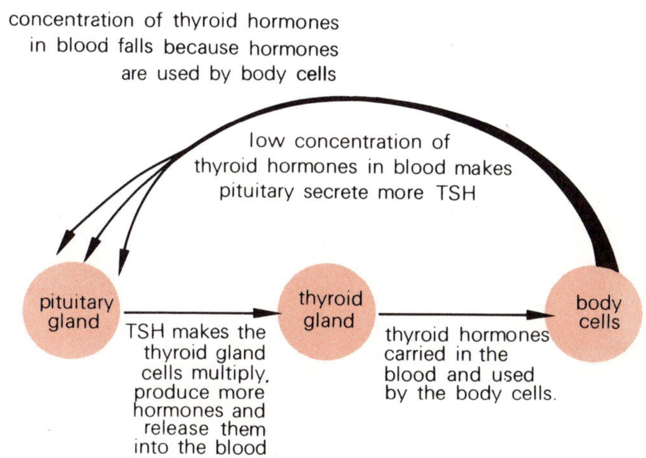

FIGURE 6·9

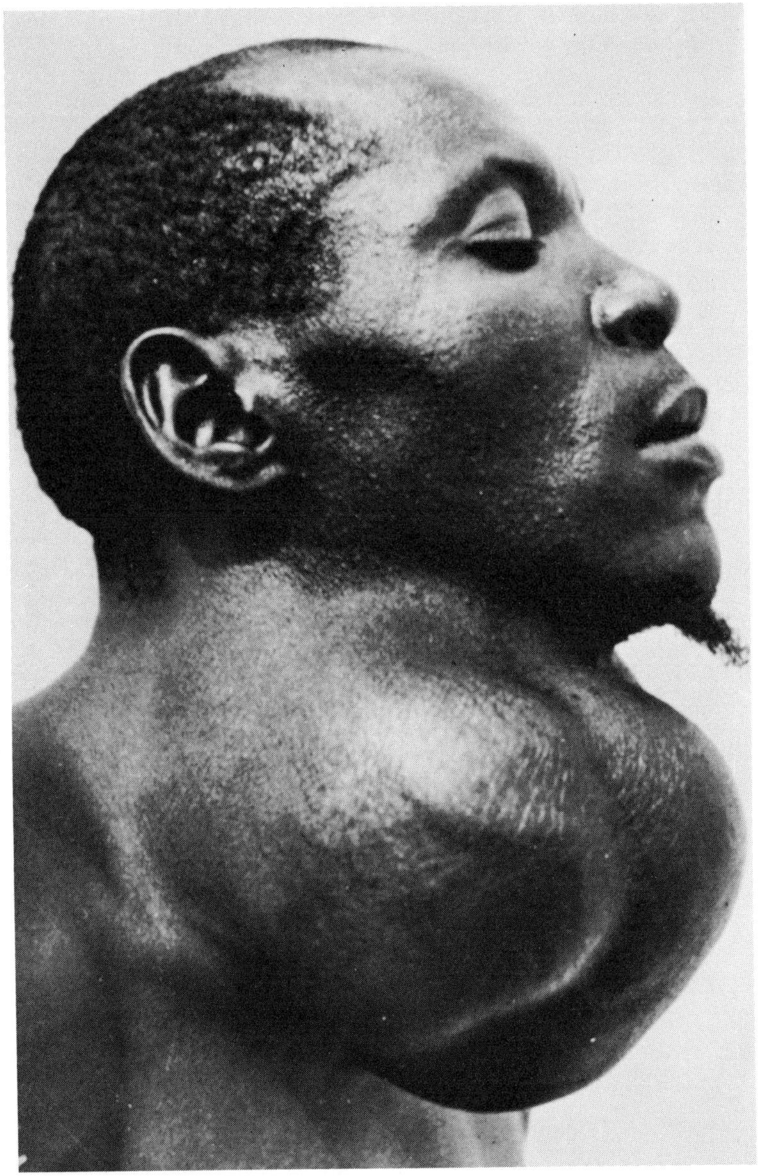

FIGURE 6·10

In some districts remote from the sea, such as the Sudan, large sections of the population develop very swollen thyroid glands (FIGURE 6.10). This gives the whole neck a swollen appearance, called *goitre*.

It is thought that the reason for goitre is lack of iodine in the diet, and it can be cured by the consumption of small quantities of iodine compounds.

A What effect would you expect lack of iodine in the diet to have on the ability of the thyroid gland to manufacture thyroid hormones?

B From FIGURE 6.9, **explain how a lack of iodine in the diet might result in enlargement of the thyroid gland.**

III

The thyroid hormones have a basic effect on body cells in that they strongly influence basal metabolic rate. They also have profound effects on growth, and these seem to be independent of the effect on basal metabolic rate.

If the thyroid of a young human does not secrete the proper quantity of hormones, the individual develops into a cretin: he remains small–a 40-year old cretin may be only 1·5 m tall–he has poor teeth and a pot belly, because of weak abdominal muscles. The reproductive organs remain infantile; the person has a vacant look and is severely mentally retarded. A 40-year old cretin may only be able to say the few words that an eighteen-month baby can formulate, and his behaviour is much the same as that of such a young child.

How the thyroid hormones have such widespread and fundamental effects is not known. From the description of the symptoms you can see that to answer such a question, or series of questions, requires extensive knowledge of how almost all the parts of the body grow and adjust themselves into a working whole. Today we are on the verge of understanding such fundamental things in biology.

All higher vertebrates (i.e. most fish, all amphibians, reptiles, birds and mammals) have thyroid glands. In amphibians the thyroid hormones control the development of the tadpole into the frog (metamorphosis). We shall now look at some of the facts that are known about this process, and they may raise ideas as to how the human thyroid controls human development.

FIGURE 6.11 shows some of the stages of development from tadpole to frog.

A **Make a list of the differences you can see between the five stages** A–E.

B **Some of the changes are internal and cannot be seen in the drawings. These include disappearance of the tadpole's frilled lips and horny jaws; shortening of the gut, associated with the change from herbivorous to carnivorous diet; development of the lungs and the air-pump mechanism.**

During metamorphosis the animal changes its method of locomotion, method of breathing, method of eating and its diet. Explain how the timing **of the changes you can see in the drawings, the changes listed above and any other changes you know of, fit in with these changes in the animal's way of life.**

C I **If the thyroid gland is removed from tadpoles by surgery at an early stage of metamorphosis, they either do not start to metamorphose, or, if they have already started, they go no further.**

II **If animals treated as in I have thyroid hormones added to the water in which they are living, they start, or continue, to metamorphose.**

III **If thyroid hormones are added to the water in which untreated tadpoles are living, metamorphosis proceeds more rapidly than normal.**

i **What do facts I and II reveal about the relationship between thyroid hormones and metamorphosis?**

ii **What additional information does fact III give you about this relationship?**

D Rana temporaria, **the common frog in Britain whose spawn you have most likely watched at some time, takes about nine weeks to pass from hatching to growth of the hind limbs to a length equal to that of the animal's torso. The tadpole is still moving in a fish-like way by means of its tail at this time. Development this far is called** prometamorphosis.

The final stages, including emergence of the fore limbs, re-absorption of the tail and many of the other changes listed above, take only a further three weeks. This development, taking less time but resulting in more profound changes, is called metamorphic climax.

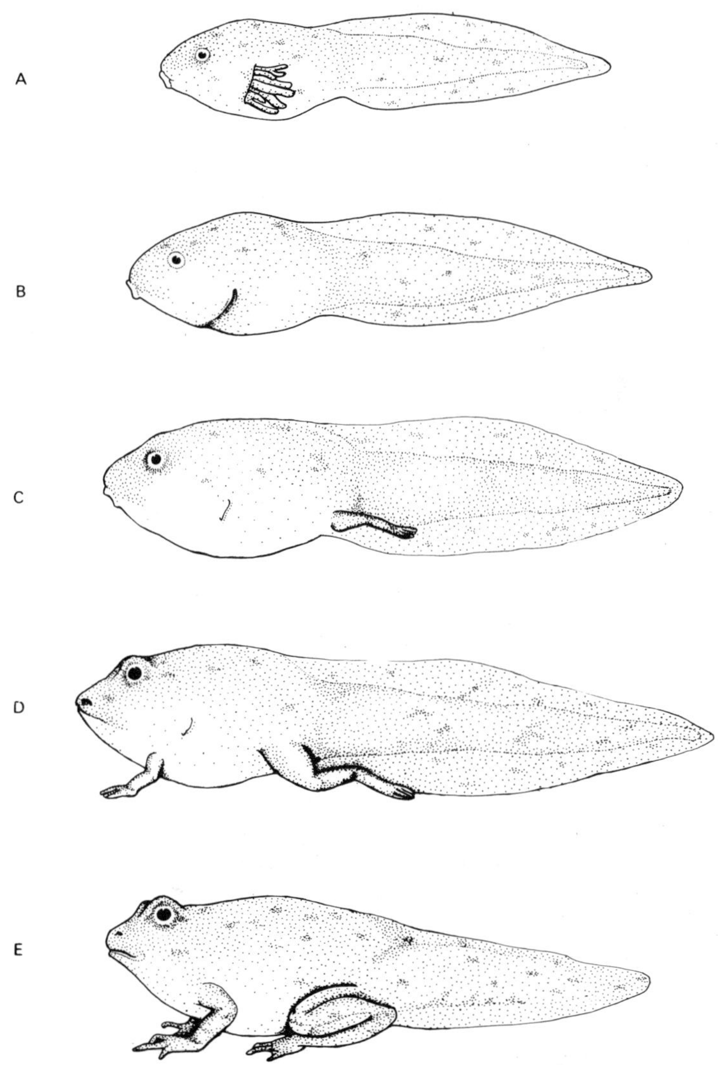

FIGURE 6·11

How do these patterns of change fit in with the observation that in very young tadpoles the thyroid gland is very small, during prometamorphosis it grows rapidly, and at metamorphic climax it is very large?

112

A **Three experiments were conducted on tadpoles of another species of frog,** *Rana cantabrigensis.* **Their thyroid glands were removed by surgery at an early stage, just after the disappearance of the external gills.**

Experiment A **Five parts per billion of thyroid hormone were added to the water in which the tadpoles were living. Metamorphosis followed this course:**

a **appearance of the forelegs behind a layer of skin;**

b **emergence of the forelegs;**

c **loss of the horny jaws;**

d **beginning and completion of the absorption of the tail. That is to say, metamorphosis followed the same course as in untreated tadpoles except that it took 40 days compared with the 30 days taken by untreated tadpoles.**

What would you infer had happened to the concentration of thyroid hormones in the untreated **tadpoles during their metamorphosis?**

B *Experiment B* **100 parts per million of thyroid hormone was added to the water in which a second batch of tadpoles were living. Treatment was started at the same stage as before, just after the removal of their thyroid glands. The course followed in metamorphosis was:**

a **loss of the horny jaws;**

b **appearance of the forelegs behind a layer of skin + rapid emergence of the forelegs + beginning of reabsorption of the tail;**

c **completion of reabsorption of the tail.**

What effect has this dose of thyroid hormone had on metamorphosis of these tadpoles?

C **It seems that for normal metamorphosis it is necessary for the concentration of thyroid hormone in the animal to increase as the process develops; but also the** rate **of this increase must be right; if the increase is too great, the normal pattern of development is disturbed.**

In *experiment C,* **a third batch of tadpoles, again at the same stage of development as before, were given gradually increasing doses of thyroid hormones in the water in which they were living. For the first five days they were given one part per billion; for the next 12 days, five parts per billion; then, when metamorphic climax was reached, increasing doses up to 1000 parts per billion. Metamorphosis followed the normal pattern and took thirty days.**

How do the results of this experiment confirm the inference made from the observed increase in size of the thyroid during metamorphosis (question 111 D)?

113

So far in this chapter we have seen examples of hormones which are produced by definite structures called endocrine glands (e.g. by the island cells of the pancreas, by the thyroid gland and by the pituitary gland), and one example of a hormone produced by cells not collected together to form a gland and which have other functions beside that of producing the hormone (i.e. by the epithelial cells of the small intestine).

We will look at another different site of hormone production, namely nerve endings.

In 1921 a German biologist called Loewi investigated factors affecting the rate of heartbeat of the frog. An isolated frog heart will continue to beat provided it is kept in a solution which contains the right proportions of Na^+, K^+ and Ca^{2+} and is oxygenated.

The heart receives two sets of nerves, but continues to beat after both sets have been cut: it has an inherent rhythm of its own. The function of the nerves is to increase or decrease the rate of its inherent rhythm.

Loewi washed the frog's heart several times to remove all remaining blood. Then he allowed it to beat for a while in the correct liquid medium. Call this medium solution I.

He then transferred the beating heart to a fresh similar solution and applied an electrical stimulus several times to one of the nerves which serves the heart, the vagus nerve. This produced the well-known effect of making the rate of heartbeat slower and decreasing the strength of each beat. He removed the heart from this liquid, which we will call solution II.

The heart was again placed in a fresh similar solution, but the vagus nerve was not stimulated. When it had reached its normal rate of beat he immersed it first in solution I, then in solution II. Solution I had no effect on the rate or strength of heartbeat. Solution II caused the rate of beat to slow down and reduced the strength of each beat.

A **What seems to have happened to solution** II **that has made it different from solution** I **in its effect on the heart?**

B **What has caused this to happen to solution** II **?**

C **When the heart was replaced in solution** II, **the rate and strength of beat diminished. What does this indicate about the possible way in which the vagus nerve influences the rate and strength of heartbeat? Explain your answer.**

D **A chemical substance called atropine has long been known to stimulate an increase in the rate of heartbeat because it abolishes the effects of the vagus nerve. When atropine was added to solution** II, **the solution no longer had any effect in slowing down the rate or strength of beat of the recovered heart. How can you explain this result?**

114

The central part of the adrenal gland (the adrenal medulla) secretes adrenalin. This hormone produces many diverse effects on the body which, collectively, prepare it for 'fight or flight'.

A **Say how the effects of adrenalin on each of the following parts of the body contributes to the 'fight or flight' condition:**

1 **the heart;**
2 **the liver;**
3 **the spleen;**
4 **the lungs;**
5 **the arteries leading to the heart and to the muscles which move the body;**
6 **the arteries leading to the skin and the gut;**
7 **the muscle of the gut wall, including the sphincters (valves) between the stomach and the duodenum, between the small and large intestine, and at the anus.**

115

A i **Make a list of the hormones you have learned about in this chapter.**

ii **Say in each case from what structure they are released.**

iii **What in general may be said about the origin of hormones?**

B **Of the hormones in your list, which produce** specific **single effects on other parts of the body and which have general effects throughout the body? Some are difficult to classify. Say which they are and why they are difficult to classify.**

C **The mammal has two methods of coordinating the functions of its body, by means of nerves and by means of hormones. Look at the following list of functions of the nervous system and say for each point how hormone coordination differs from nervous coordination.**

i **Nervous control is rapid. Nerve impulses travel at a speed of between 1 and 120 metres per second.**

ii **The nervous system receives information about the environment from the sense organs.**

iii **Nerve impulses cause activity in muscles and in glands.**

iv **Sometimes nervous coordination takes place exclusively via the spinal cord (reflex action). Information about reflex activity may be sent to the brain, which may or may not exert an influence on it.**

v **Nerve influences on muscle and gland cells are very short lived.**

vi **The number of muscle cells under the control of one nerve cell varies from about 2 to about 200-300.**

vii **Nervous coordination is highly specific. The nerve cells form fine, well-defined channels of communication.**

viii **The brain and spinal cord are large organs. There is a vast number of nerve cells in the body; for example about 100 000 nerve cell axons enter and leave each hand.**

116

We have seen that some hormones are produced in definite organs called endocrine glands. It is likely that most, if not all, of the endocrine glands in the human being are known. But suppose you are a biologist investigating endocrine functions in a reptile. You discover a structure that looks like a gland and that has no known function. You suspect that it may be an endocrine gland, and carefully remove it during a surgical operation.

The animal recovers and after a while shows symptoms of disease when compared with a control animal.

A i What would the control animal be?

ii In choosing the control animal, what features would you look for?

iii What treatment would the control animal receive during the course of the experiment?

B In order to determine whether the gland you have removed secreted a hormone which prevented the symptoms of disease, you could grind it with sand to make an extract, filter the extract, and inject it into the animal. Bayliss and Starling did this when they were investigating secretin in the pancreas.

Suppose this treatment did not work. You could not take such evidence as proof that the gland did not secrete a hormone during life. Why not?

C Suppose, on the other hand, that the extraction–injection treatment did work, and that the disease symptoms were relieved. This would be good evidence in favour of the idea that the gland produced a hormone in life which is released into the bloodstream, preventing the disease symptoms, but it would not be proof. Why not?

D An extract of the gland would contain many chemical substances. What, then, would be a good next step in your investigation?

E Suppose you had a large number of these reptiles and some of them showed disease symptoms similar to those produced by surgical removal of the gland you had discovered. Describe two distinct ways in which you could use the diseased animals in support of your hypothesis that the gland produced a hormone which normally prevented the appearance of the disease symptoms.

117

FIGURE 7.1 shows three lumbar vertebrae of a rabbit. A was soaked for 24 hours in 20% hydrochloric acid. This dissolves away all the calcium salts that the bone contains. B was heated strongly until no more smoke came off. C was untreated.

A **The bones were not, of course, identical, but leaving aside such small differences as you would expect between different lumbar vertebrae (including overall size), what was the effect of**

i **steeping in acid**

ii **heating on the shape of the bones A and B ?**

B **Steeping in hydrochloric acid removed the inorganic matter from the bone; heating removed the organic matter. What does the answer to A tell you about the evenness with which the inorganic and organic matters are disposed in the bone?**

C **After treatment, bone A was rubbery–all parts of it were soft, could be bent and returned to their former positions when the force bending them was removed.**

Bone B was hard.

What physical properties, therefore, do you think that:

i **the organic,**

ii **the inorganic material gives to the bone?**

D **Two successive lumbar vertebrae from a rabbit were dried to constant weight in a hot air oven at 100°C. Bone A weighed 1·251 g. Bone B weighed 0·998 g.**

i **Describe the technique of drying an object** to constant weight.

ii **Bone contains organic matter. What might heating to 100°C do to the bone to make the estimation of its true dry weight difficult or impossible?**

iii **Bone A was then treated with 20% hydrochloric acid for 24 hours. It was washed and dried to constant weight at 100°C. It then weighed 0·411 g.**

Calculate the percentage by weight of inorganic material in the bone.

iv **Bone B was heated fiercely till no more smoke came off; then it was cooled in a desiccator containing calcium chloride and re-weighed. Its new weight was 0·563 g. Calculate the percentage by weight of organic material in the bone.**

v **How do the results of these two experiments agree with one another? Assume that a discrepancy between results of up to 2% can be attributed to experimental error. Make three suggestions as to why the results differ by more than 2%.**

118

When we look at the microscopic structure of bone we find that

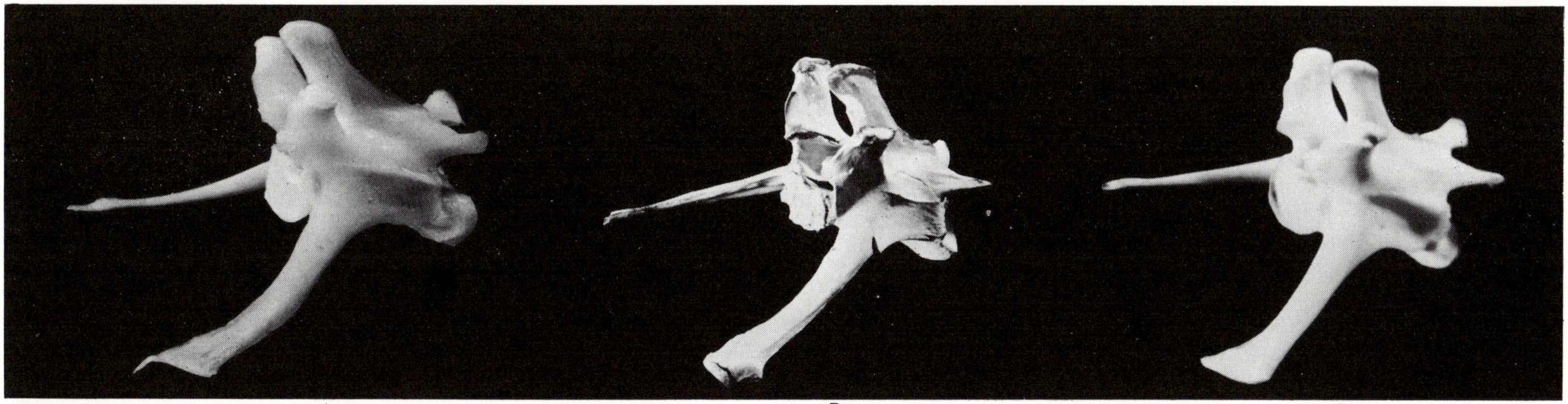

FIGURE 7·1 A B C

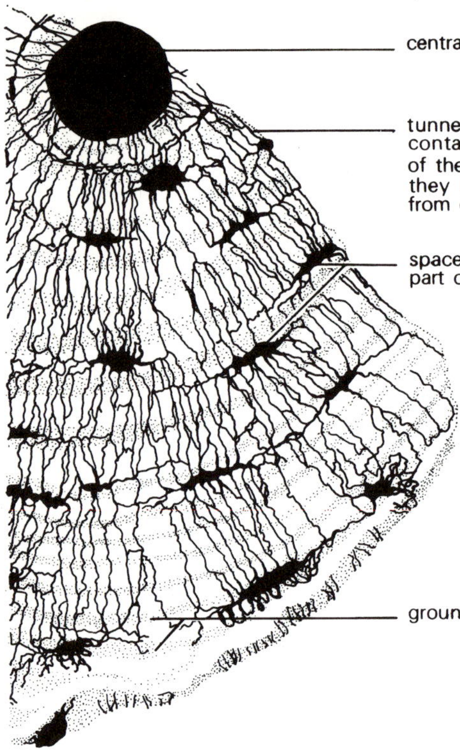

FIGURE 7·2

the living bone cell has a curious shape. There is a large central part containing the nucleus, and radiating from it in all planes very thin, branched arms of cytoplasm.

FIGURE 7.2 looks as though it shows this, but in fact the black shapes we see there are hollows in the *ground substance* in which the bone cells lie. The central part of each cell lies in a large hollow which it almost completely fills. Each fine branch of the cytoplasm runs into a tunnel in the ground substance; we can see the end of one of the tunnels, photographed with an electron microscope, in FIGURE 7.3. FIGURE 7.4 shows what a few bone cells would look like if the ground substance round them were removed.

The ground substance is a mixture of millions of interlocking tiny

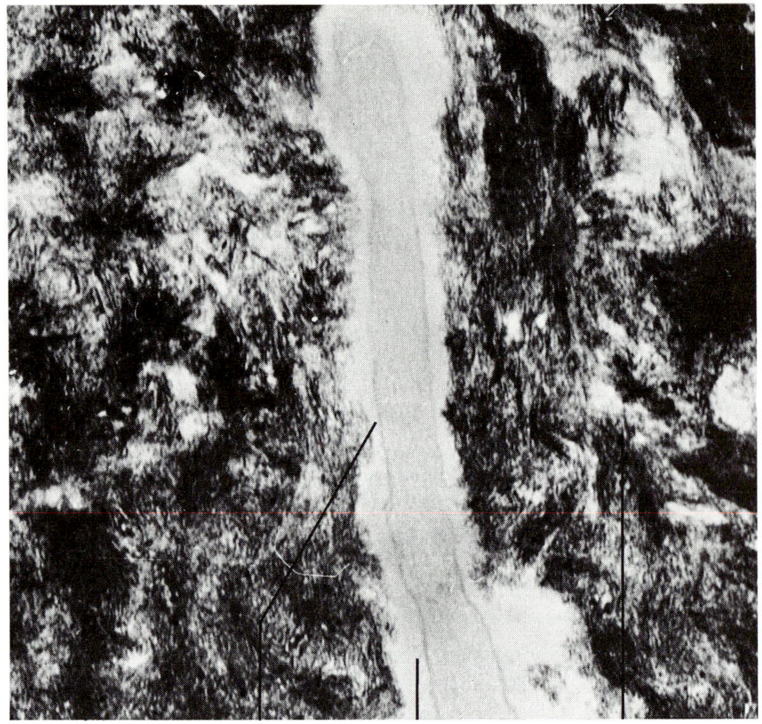

FIGURE 7·3

crystals of calcium salts (phosphate and carbonate) mixed up with millions of interlocking fibres of a tough protein.

A **From the above description, give a more accurate account of the way in which the inorganic and organic materials in bone are arranged than was possible in question 117 B.**

B **Bone, being riddled throughout by living cells, is very much a living tissue; also the non-living ground substance is in a constant state of change. If a compound containing radioactive phosphorus is eaten in the food, the radioactive material appears in the ground substance within a few hours. It is estimated that about one fifth of the ground substance of the human skeleton is changed per month.**

How does the structure of the bone cells suggest that this

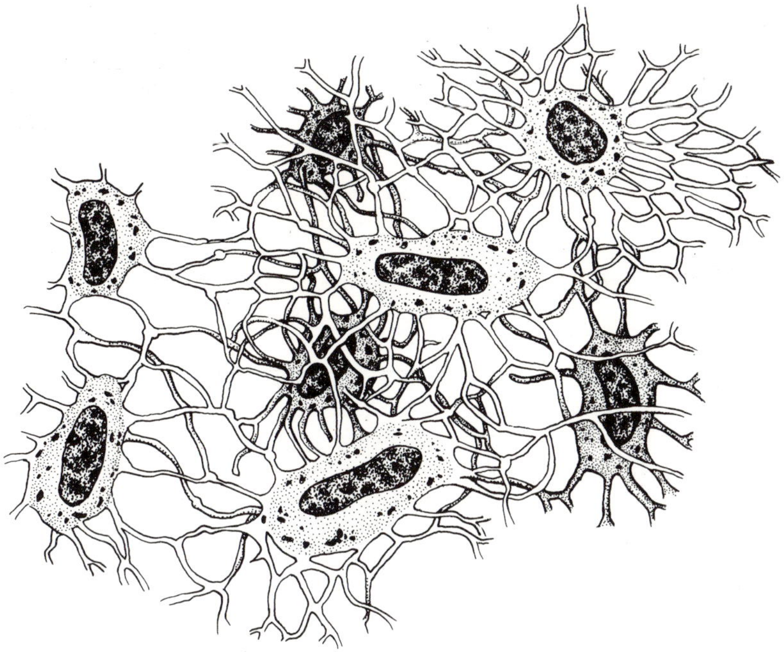

FIGURE 7·4

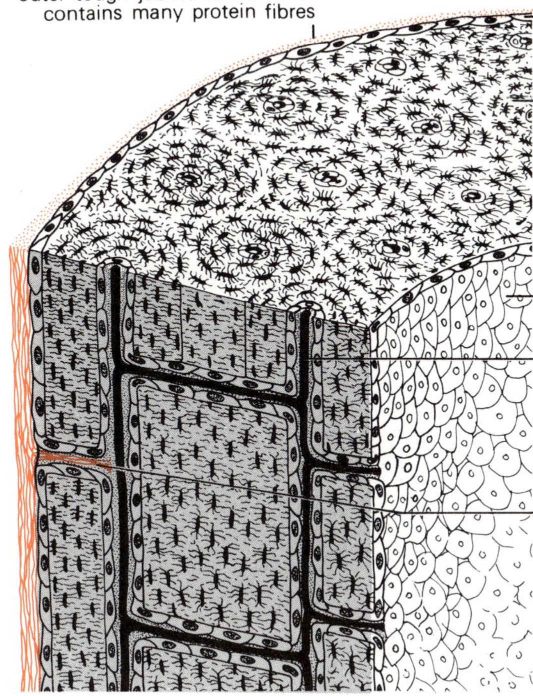

FIGURE 7·5

change in the ground substance is brought about by them?

C I The bone cells are arranged roughly in columns, forming concentric layers which surround holes (FIGURE 7.5). The holes contain arterioles, venules, capillaries and lymph vessels.

II No capillaries or lymph vessels penetrate the ground substance which, because of the large number of crystals it contains, is only slightly permeable to substances in solution.

Bearing these facts in mind, how do you suppose the radioactive phosphorus:

i reaches the ground substance,

ii becomes incorporated into it?

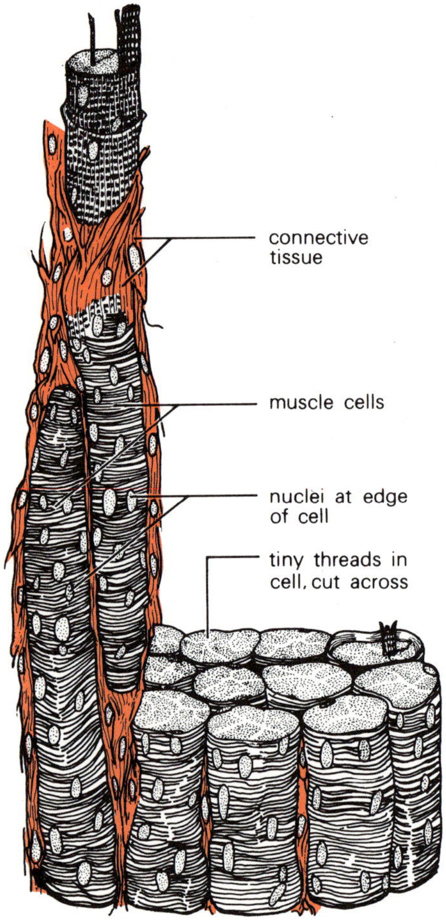

FIGURE 7·6

D **When a baby is born its bones are small. As it grows, so its bones grow. New bone is formed by a layer of cells just below the fibrous outer jacket of the bone (see** FIGURE 7·5 **).**

One might suppose that more and more layers of bone were formed as the bone grew, so that a section of it would have a layered appearance. In fact, however, at all ages from birth to maturity, a section of bone appears much like FIGURE 7·5**; one cannot make out any layers. As soon as new material is laid down, it is incorporated into the rest of the bone so that it forms an indistinguishable part of it.**

From the information you have gathered in this question, how do you suppose this incorporation takes place?

119

Muscles and bones act together to maintain the posture of the body. To understand what this means, think of the absence of posture when a person has fainted or is under anaesthetic. They also act together to produce movement. Muscles must therefore be attached to bones in some way, and in this question we are going to find out how.

The function of muscles cells is to shorten (contract), so it is no surprise that their shape is very long and narrow (FIGURE 7.6). Single cells may be many centimetres long and about 10–100 mm wide. Each cell has many nuclei, all arranged at the edge of the cell, and internally it is divided up into a large number of long, thin tiny threads. Surrounding the cell is a thin layer of *connective tissue* separating it from the next cell and holding all the cells together. The connective tissue consists mainly of a large number of very tough, unstretchable fibres of protein (collagen fibres).

FIGURE 7.7 is a transverse section of a human sartorius muscle. It is easy to see that the muscle cells are grouped together in blocks, the blocks being separated from each other by connective tissue. Smaller blocks are grouped into larger blocks, each surrounded by connective tissue; and the whole muscle is bounded on its outside by a thick layer of connective tissue.

Thus connective tissue not only surrounds each muscle cell (FIGURE 7.6) but also all the smaller and larger blocks of muscle cells. The arrangement is made clear by FIGURE 7.8, which is a drawing of a small portion of FIGURE 7.7.

A FIGURE 7.8 **shows that the connective tissue contains many blood vessels.** FIGURE 7.9 **shows some of them in more detail.**

i **Suggest a reason for the very rich supply of capillaries to the muscle cells revealed by** FIGURES 7.8 **and** 7.9.

ii **Muscle cells are arranged in small groups, or blocks. This means that the blood vessels serving the muscle**

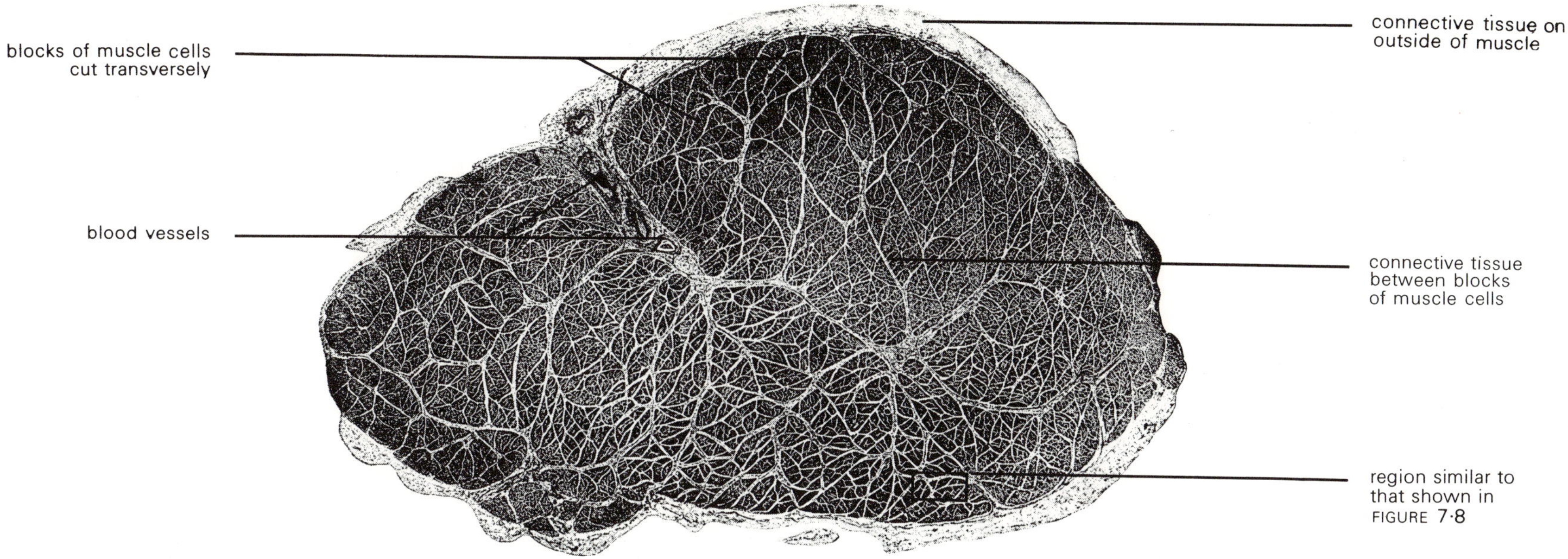

FIGURE 7·7

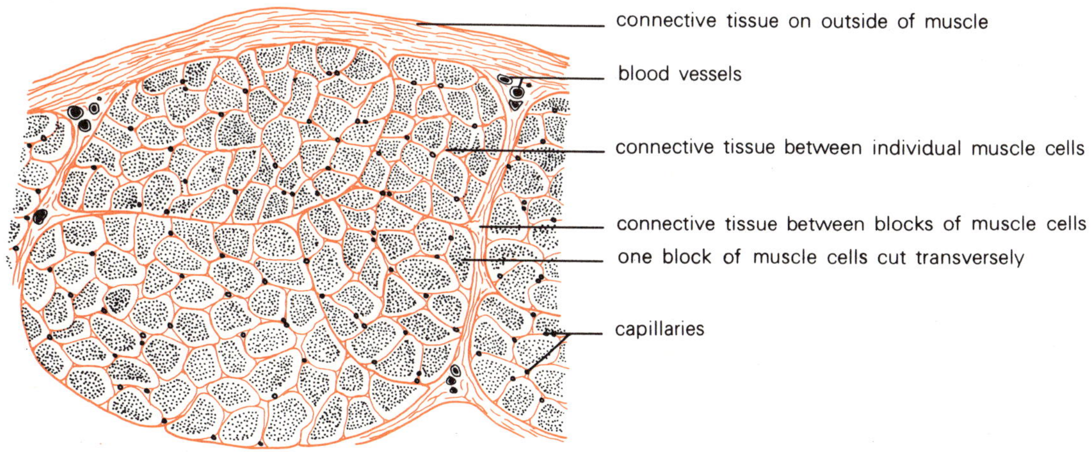

FIGURE 7·8

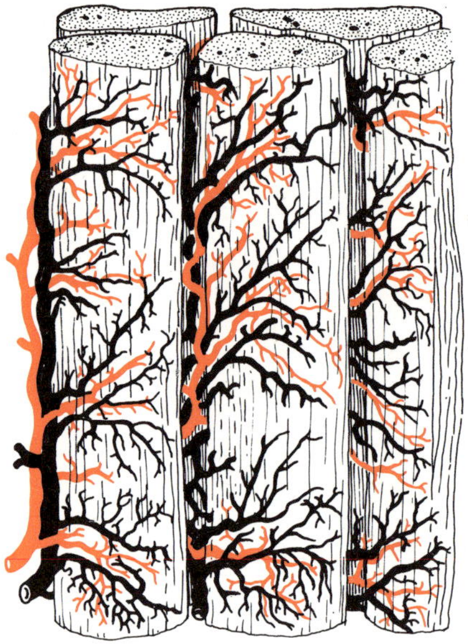

a

FIGURE 7·9a

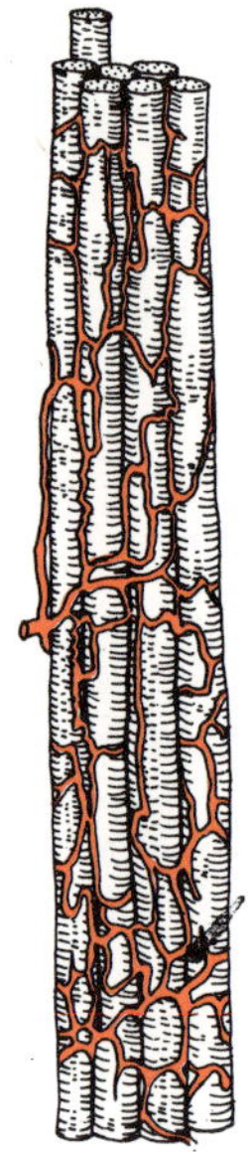

b

FIGURE 7·9b

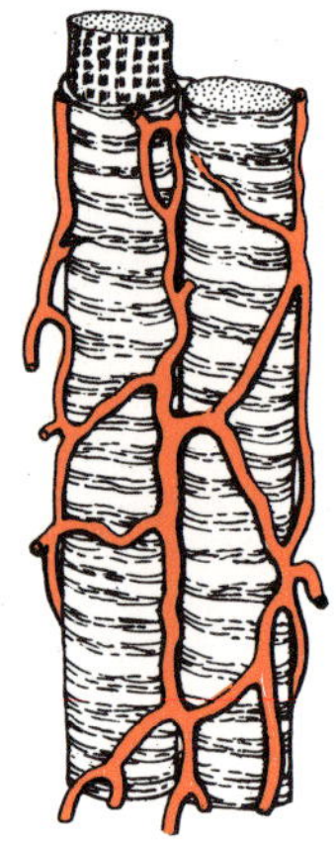

c

FIGURE 7·9c

are fairly wide when they run in between the blocks (FIGURE 7.9a) and only become narrow capillaries when they divide and run out in between individual muscle cells (FIGURE 7.9b and c).

Consider another possibility, that instead of the muscle cells being arranged in blocks, the whole sartorius muscle (for example) might consist instead of evenly-distributed muscle cells.

What would the arrangement of blood vessels be in this made-up case, and why is the arrangement which really exists better than this from the point of view of rapid delivery of blood to the muscle cells?

iii FIGURE 7.9c shows that the capillaries surrounding muscle cells are not straight but crinkled. How is this fact related to the function of the muscle cells?

B What do FIGURES 7.6, 7.7 and 7.8 tell you about the connections between the various parts of the connective tissue in the muscles? Say how the pictures give you your answer.

C The whole muscle contracts because the individual muscle cells, indeed the individual tiny threads within the muscle cells, contract. In other words, the smallest parts of the muscle cells are the active structures which shorten.

A muscle like the human sartorius (FIGURE 7.7) contains thousands or tens of thousands of muscle cells. Bearing in mind the nature of the connective tissue in the muscle, suggest how it harnesses contraction of individual cells so that the entire muscle contracts together.

D You are now in a position to see how muscles are connected to bones and how they transmit their pull to bones. Look at FIGURES 7.10a and b. These show how muscle cells connect with tendons. Tendons are made of millions of tough inelastic protein fibres. FIGURE 7.10a is a diagram, FIGURE 7.10b a photograph of the real thing.

Next look at FIGURE 7.10c (page 104, detail on page 103). This shows some of the muscles of the human thigh and how they are connected to the femur, the tibia and the fibula. Tendons are shown in red.

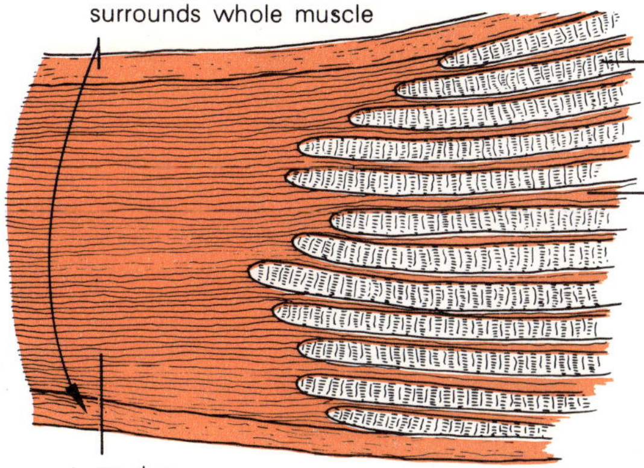

FIGURE 7·10a

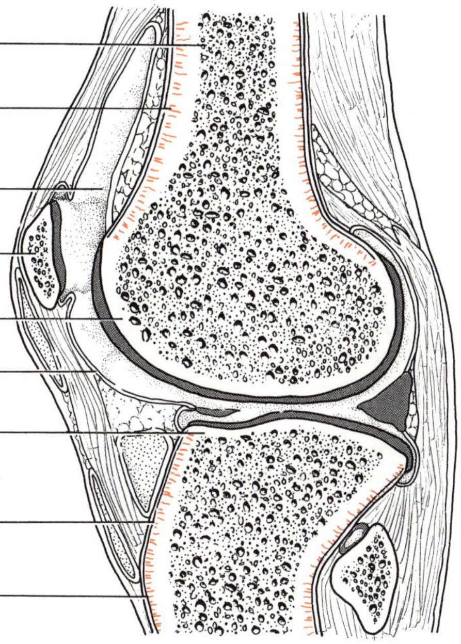

FIGURE 7·10c (detail)

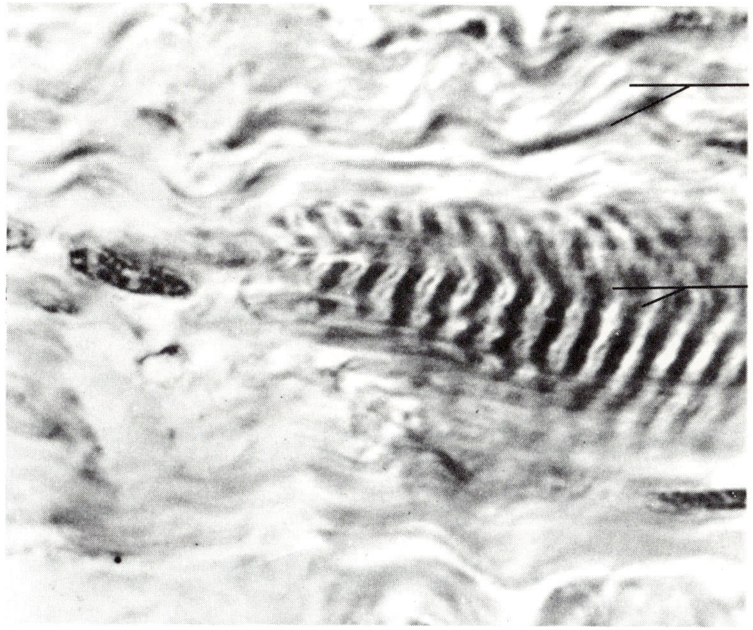

FIGURE 7·10b

Notice that surrounding the bone is a thin jacket, again containing millions of tough, inelastic protein fibres. Look back to FIGURE 7.5 **and find this jacket there, and see how it enters the bone.**

i Write as exact a description as you can of the way in which the rectus femoris and vastus lateralis muscles are attached to the tibia. State in your description how any of the pictures referred to in this question illustrate your answer.

ii From FIGURES 7.10a **and** b, **describe how muscle cells end in a tendon.**

iii From FIGURE 7.10b, **what can you say about the nature of the fibres which make up the tendon?**

iv Imagine the leg raised and the knee bent. The knee is now straightened by contraction of the rectus femoris and vastus lateralis muscles. What part do each of the tendons labelled X, Y **and** Z **in** FIGURE 7.10c **play in bringing about this movement?**

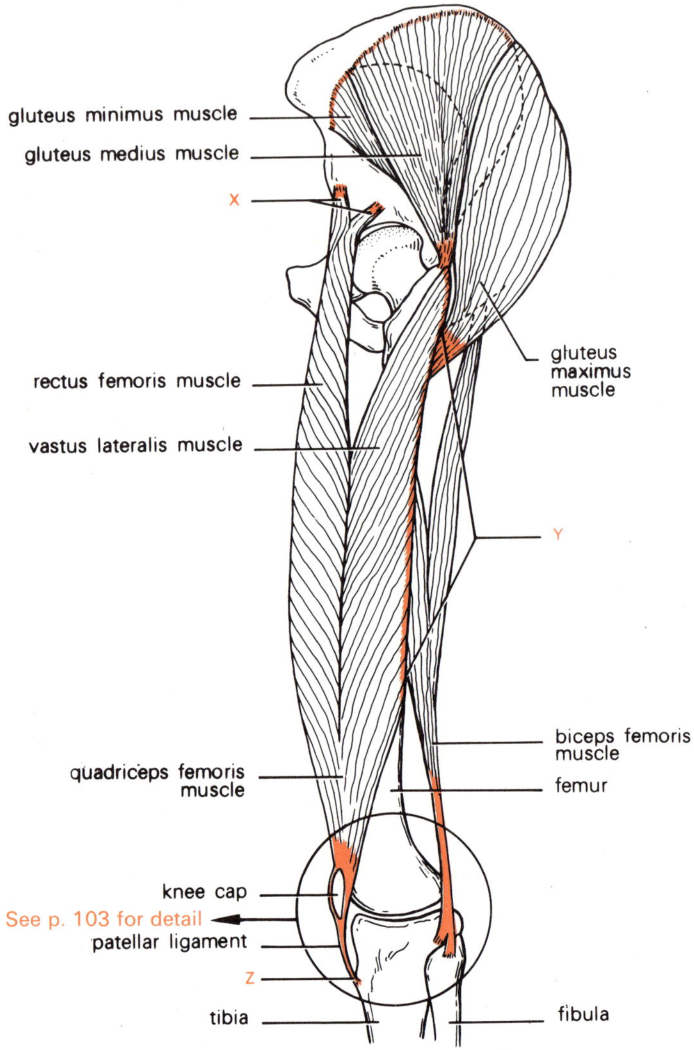

FIGURE 7·10c

120

We are going to discover something about the way in which muscle contracts. We shall look only at *striped* muscle, the muscle which moves bones, constitutes the bulk of the arms, legs and trunk, and forms most of the meat which we eat.

We have seen something of its structure in FIGURES 7.6, 7.7 and 7.8. The cells are very long, have many nuclei, contain long tiny threads, and are surrounded by connective tissue which transmits the force of their contraction to tendon and thence to bones.

FIGURE 7.11 shows us more detail of the inside of a muscle cell. The long thin threads which the cell contains are called *myofibrils*. FIGURE 7.11c shows a group of myofibrils pulled out of a muscle cell; FIGURE 7.11d shows part of one myofibril.

FIGURE 7.12a is a photograph of a longitudinal section of part of three muscle cells. It corresponds to FIGURE 7.11c. FIGURE 7.12b is an electron micrograph of part of a longitudinal section of two myofibrils. It corresponds to FIGURE 7.11d.

A **These figures show that the stained myofibril has a banded appearance. The higher the magnification, the greater the detail that can be seen inside the bands (see** FIGURES 7.11d **and** 7.12b)**; but at a low magnification** (FIGURE 7.12a) **there appear to be only two kinds of band, one stained dark, the other unstained.**

How are the dark and light bands of the various myofibrils arranged with respect to each other?

B **When we look at** FIGURES 7.11d **and** 7.12b **we see:**

1 **that the light** I **band is crossed by a dark** Z **band;**

2 **that the dark** A **band has in its centre a light region called the** H **band;**

3 **that the** H **band has a dark** M **band crossing it.**

The separate myofibrils in the muscle cell are attached to one another by connections between the M **bands. What relevance has this fact to your answer to question** A **? Bear in mind that when the muscle cell contracts, all the myofibrils within it contract at the same time.**

C FIGURE 7.12a **shows that the** I **and** A **bands (light and dark bands) alternate.** FIGURE 7.12b **shows a regular recurring pattern of** I **and** A **bands, so if we look at one section of the bands in detail, we shall know the structure of the whole myofibril.**

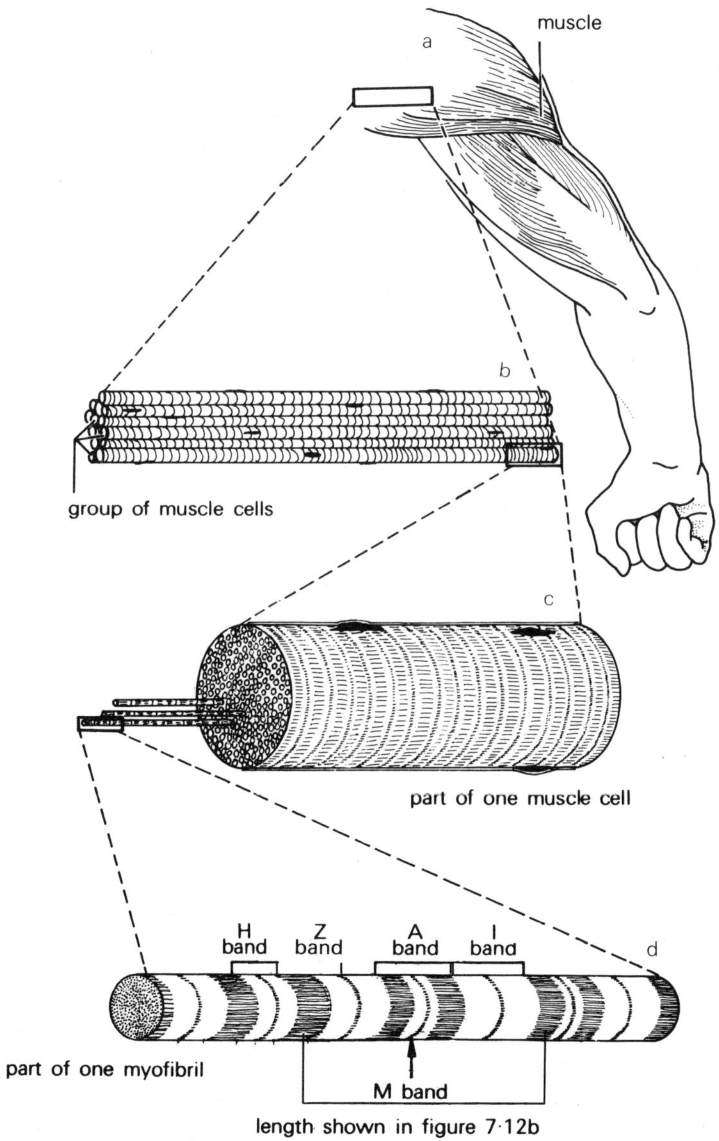

FIGURE 7·11

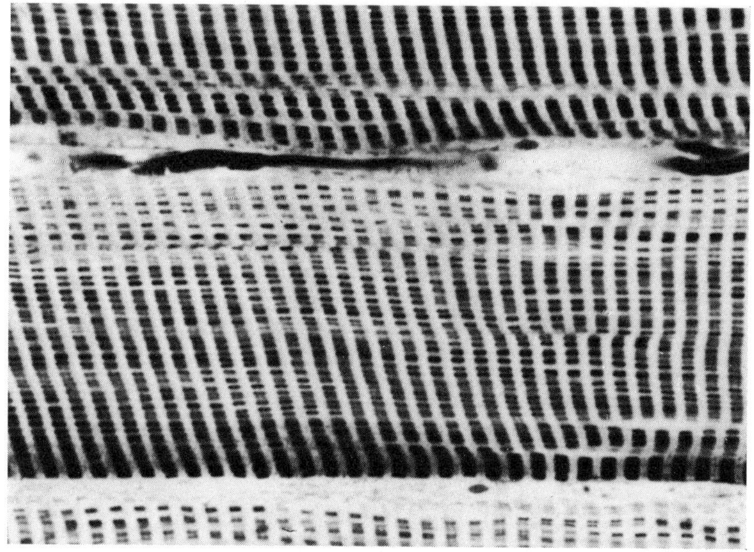

FIGURE 7·12a

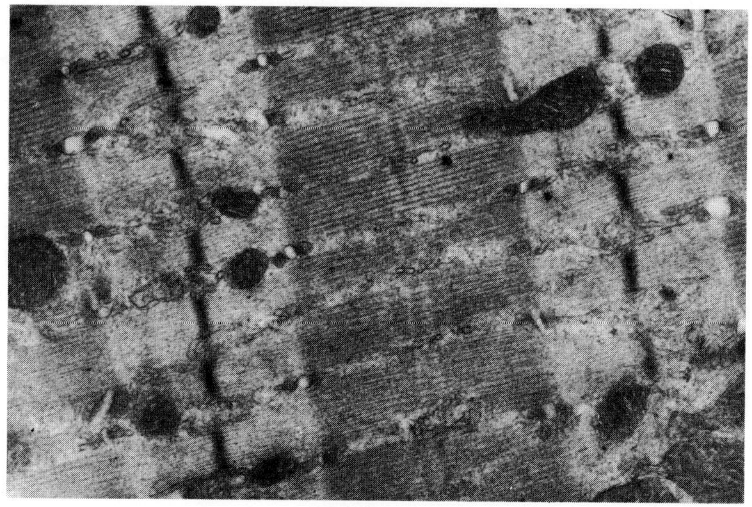

FIGURE 7·12b

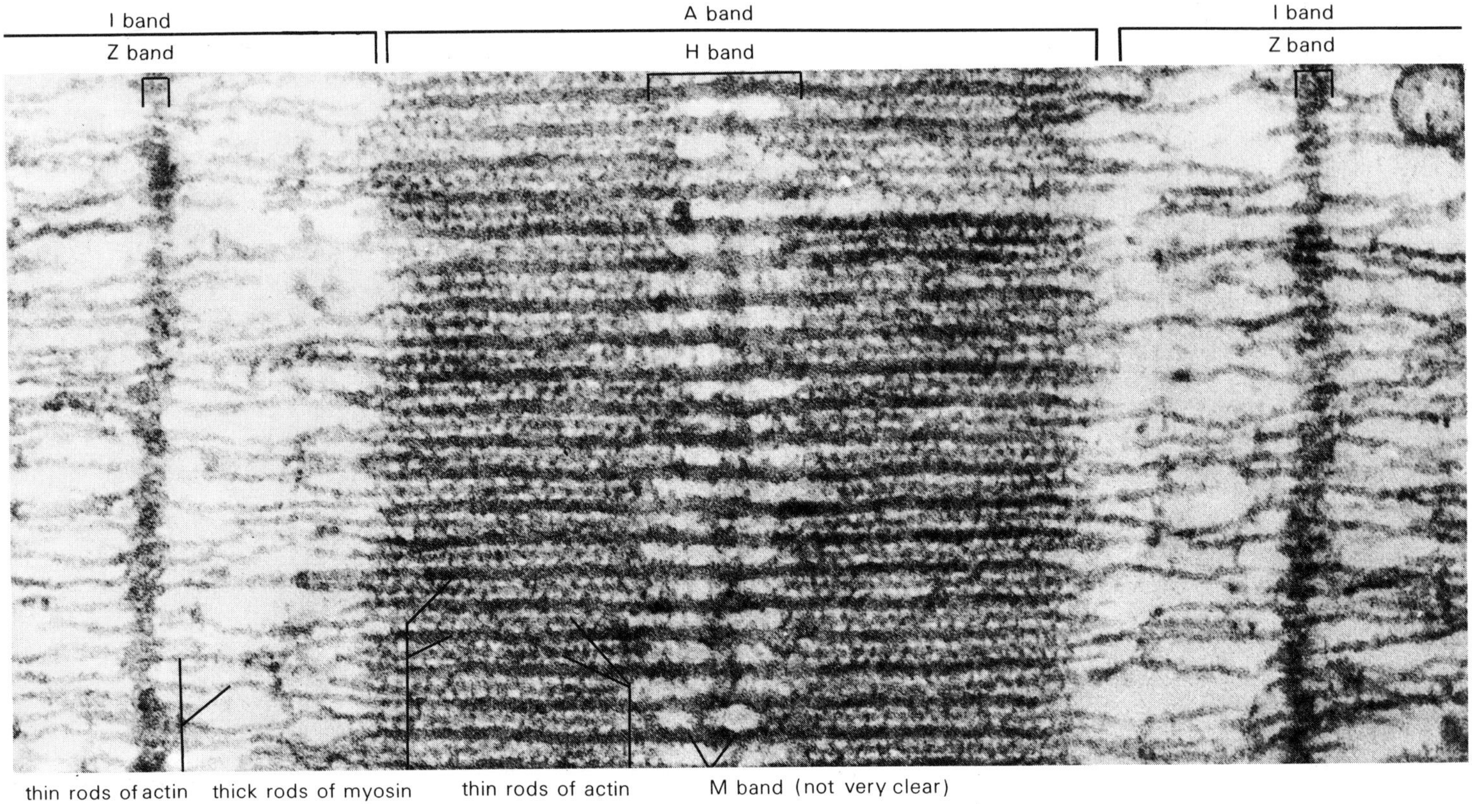

FIGURE 7·13

We can see this detail in FIGURE 7.13, **which is an electron micrograph of a longitudinal section through one** A **band and part of two** I **bands.**

Notice that the A **band contains thick rods. They consist of a protein called** myosin. **Between each pair of thick rods lies a thin rod, which is made of protein called** actin.

From FIGURE 7.13, **what becomes of the actin rods:**

i **in the** I **band,**

ii **in the** H **band?**

D **When a muscle cell contracts, all the myofibrils within it contract. The widths of** some **of the bands in the myofibrils change. This is shown diagrammatically in** FIGURE 7.14.

a **shows part of a relaxed myofibril;**

b **part of a partly contracted myofibril;**

c **part of a fully contracted myofibril.**

All the other parts of the myofibril (the repeating pattern of dark and light I **and** A **bands) change in the same way at the same time.**

i **What has happened to the width of the** A **band during contraction (measure it with a mm ruler)?**

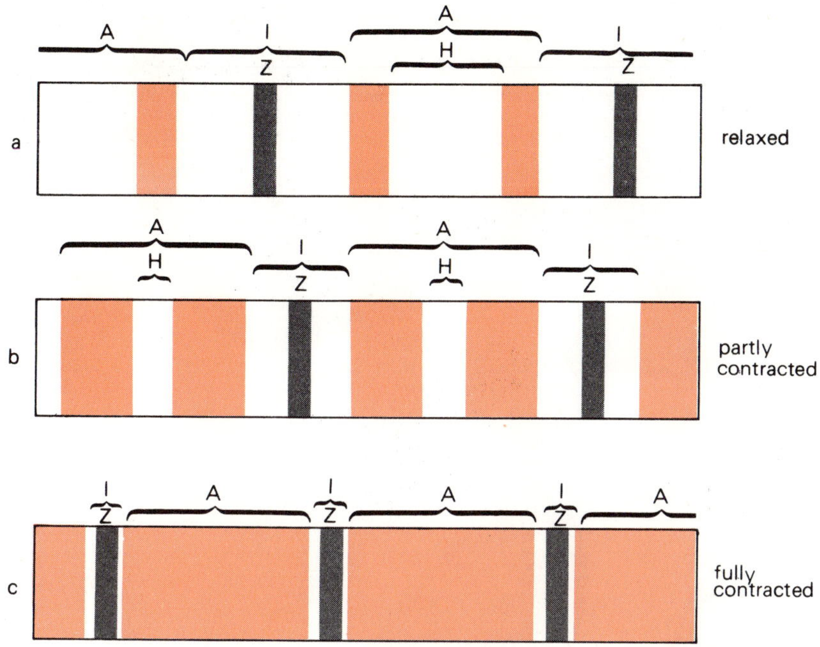

FIGURE 7·14

ii What has happened to the width of the I band during contraction?

iii What has happened to the width of the H band during contraction?

E Summarizing your answer to D, how has the myofibril shortened during contraction?

F FIGURE 7.15 **is a diagram of the arrangement of actin and myosin rods in the** I, A **and** H **bands in part of a myofibril.**

7.15b **corresponds fairly closely to the electron micrograph shown in** FIGURE 7.13. 7.15a **shows the arrangement of actin and myosin rods when the myofibril is relaxed as in** 7.14a. 7.15b **shows the arrangement when the myofibril is partly contracted, as in** FIGURE 7.14b. FIGURE 7.15c **shows it when the myofibril is fully contracted, as in** FIGURE 7.14c.

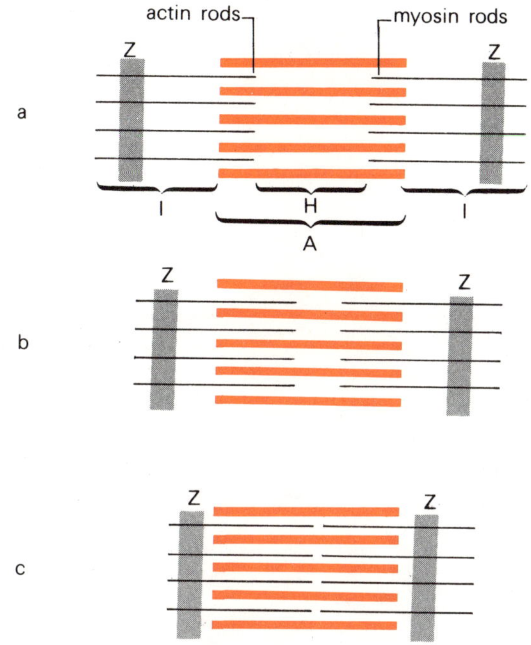

FIGURE 7·15

Considering FIGURE 7.15, **how do you suppose the change of width of the** I **and** H **bands during contraction are brought about? In other words, what is the change which takes place when myofibrils, and hence muscle cells, contract?**

G From FIGURE 7.15, **what factor do you suppose will determine the maximum amount that a myofibril can contract?**

121

FIGURE 7.12b shows that there are mitochondria packed between the myofibrils. A very small part of a muscle cell is shown in this photograph, and there is a very large number of mitochondria in a whole muscle cell.

What is the significance of this?

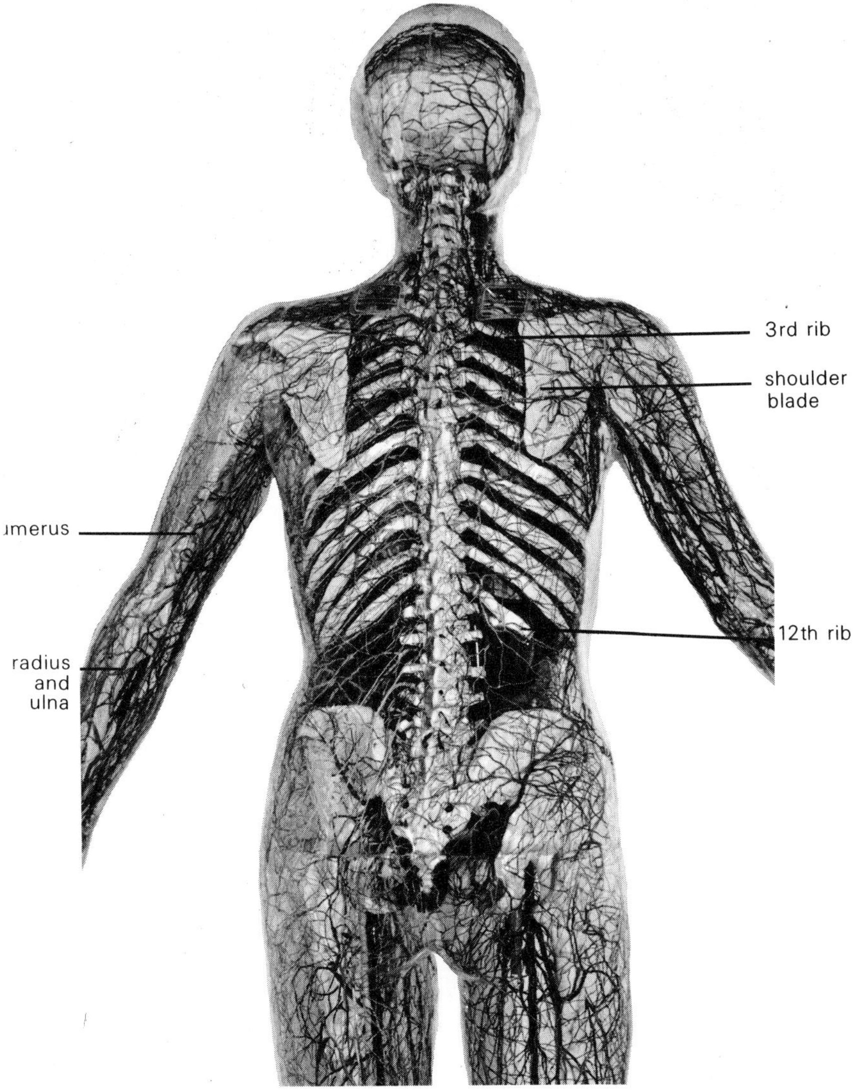

FIGURE 7·16

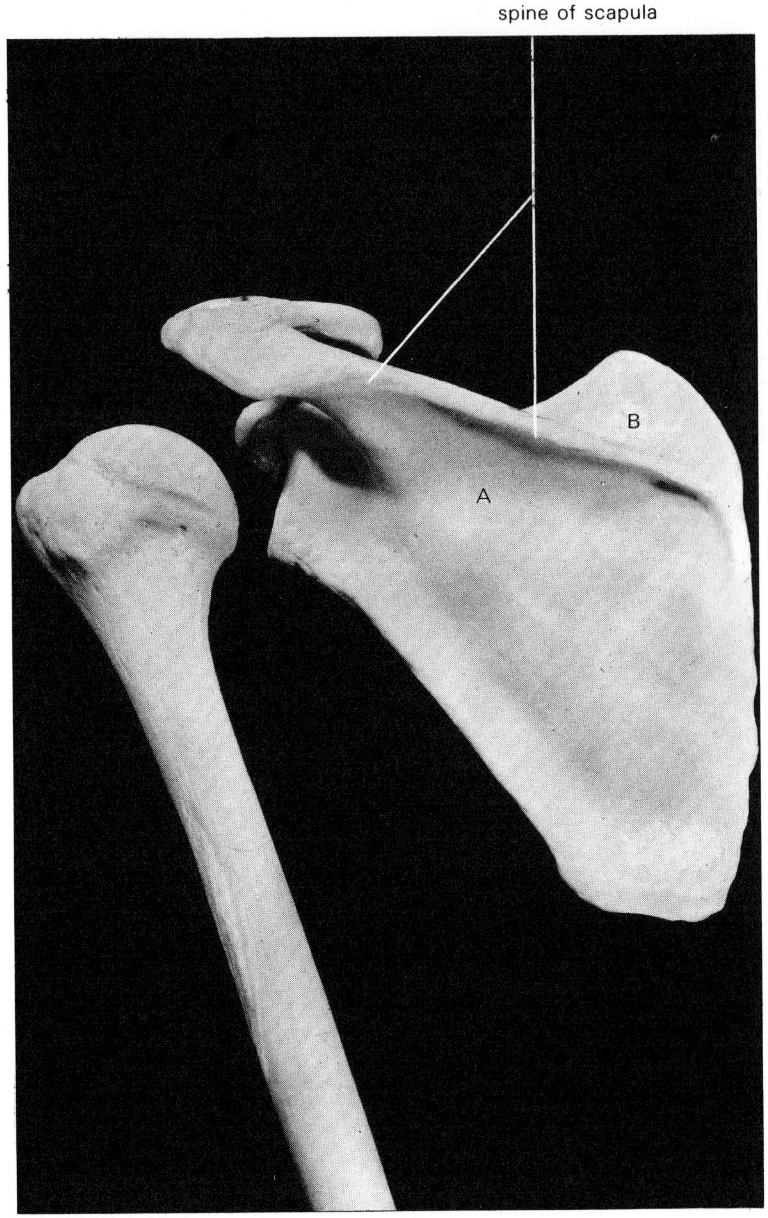

FIGURE 7·17

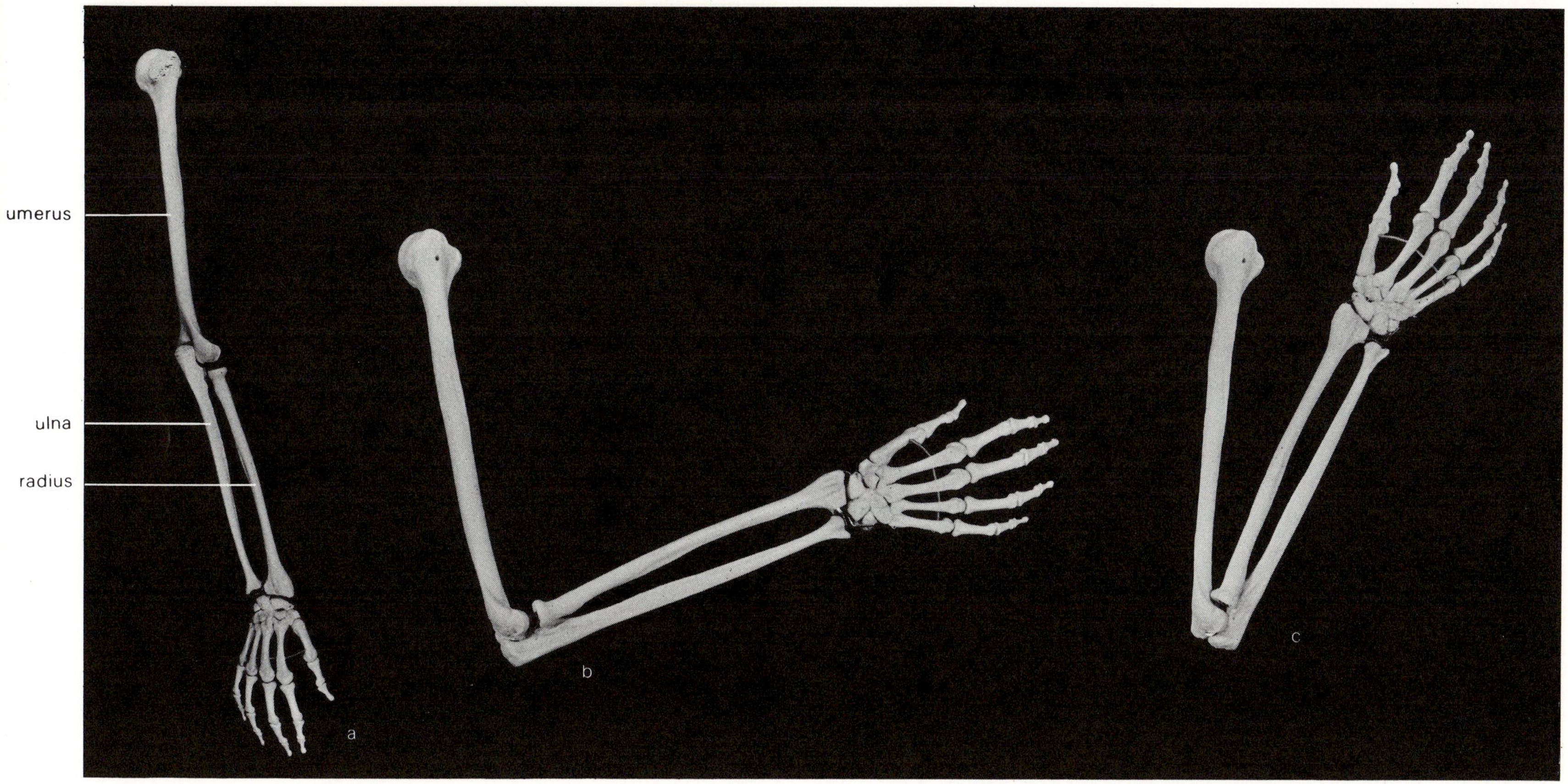

FIGURE 7·18a, b & c

122

FIGURE 7.16 is a photograph of a 'transparent woman' model from behind. You can see in it the positions of the shoulder blades (scapulae) and the arm bones. FIGURE 7.17 is a photograph of a human scapula and the upper part of the humerus.

A **From FIGURE 7.16, write an accurate description of the position of the scapulae.**

B **Is the scapula shown in FIGURE 7.17 from the right or the left side of the skeleton? Give the reasons for your answer.**

C **FIGURE 7.18a, *b* and c show three positions of the human arm bones. In a the arm is fully extended; in b the forearm is half raised; in c it is fully raised.**

i **Are these bones from the right or the left side of the skeleton? Give two reasons for your answer, one associated with the humerus, the other with the hand bones.**

ii **What sort of movement is possible at the elbow joint?**

iii **In FIGURE 7.18a, the forearm is shown fully extended. What structures limit this extension? Describe their appearance in FIGURE 7.18a.**

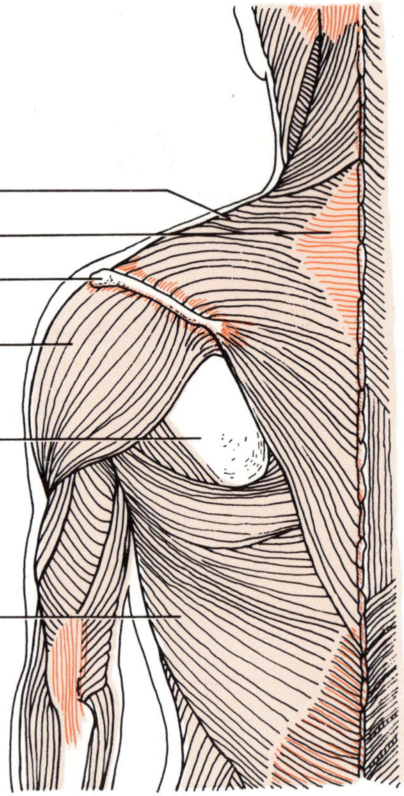

FIGURE 7·19

D A large number of muscles have their tendons attached to the shoulder blade. Two of them, both attached to the spine of the shoulder blade, are shown in FIGURE 7.19, **the trapezius and the deltoid.**

The deltoid is attached to the outside of the humerus and brings about various movements of the arm. To do this it must have something steady to pull against.

i To which side of the spine A **or** B **of the shoulder blade shown in** FIGURE 7.17 **is the deltoid attached?**

ii From FIGURE 7.19, **describe the attachments of the trapezius muscle.**

iii What important contribution do you think the trapezius makes to the function of the deltoid?

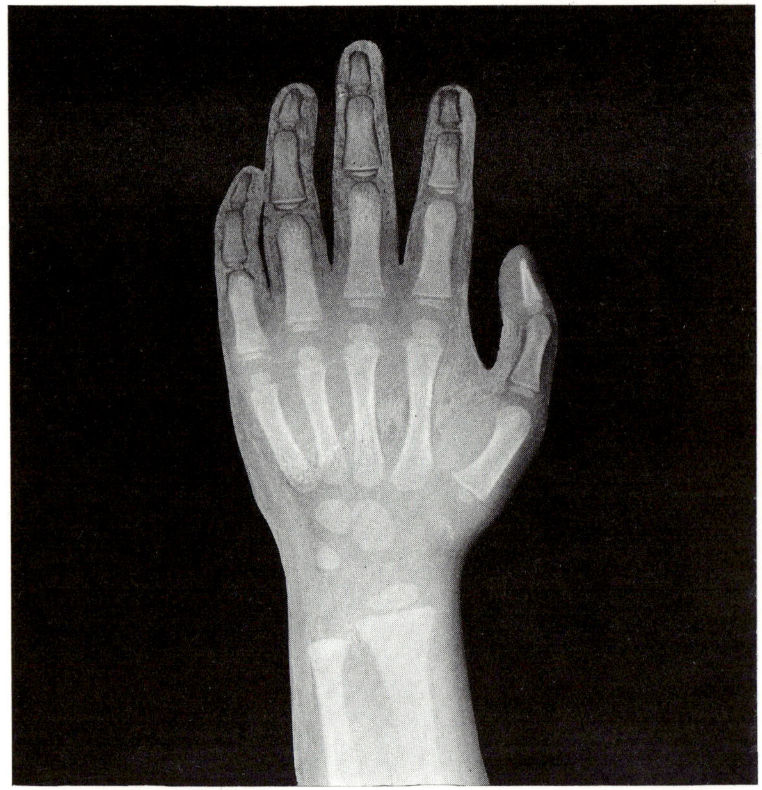

FIGURE 7·20a

123

FIGURE 7.20b is an X-ray of an adult hand. It looks the same as FIGURE 7.18, but we can see here how the arm and wrist bones articulate in the living person.

FIGURE 7.21 shows how the radius is bound close to the ulna by a bandage-like ligament.

A Describe the ends of the radius and the ulna near the wrist in FIGURE 7.20b.

B From FIGURE 7.20b, **describe the articulation between the arm bones and the wrist bones.**

C Now lay your hand with its back on a table. Turn it over, so that its front lies flat on the table. From the X-ray in

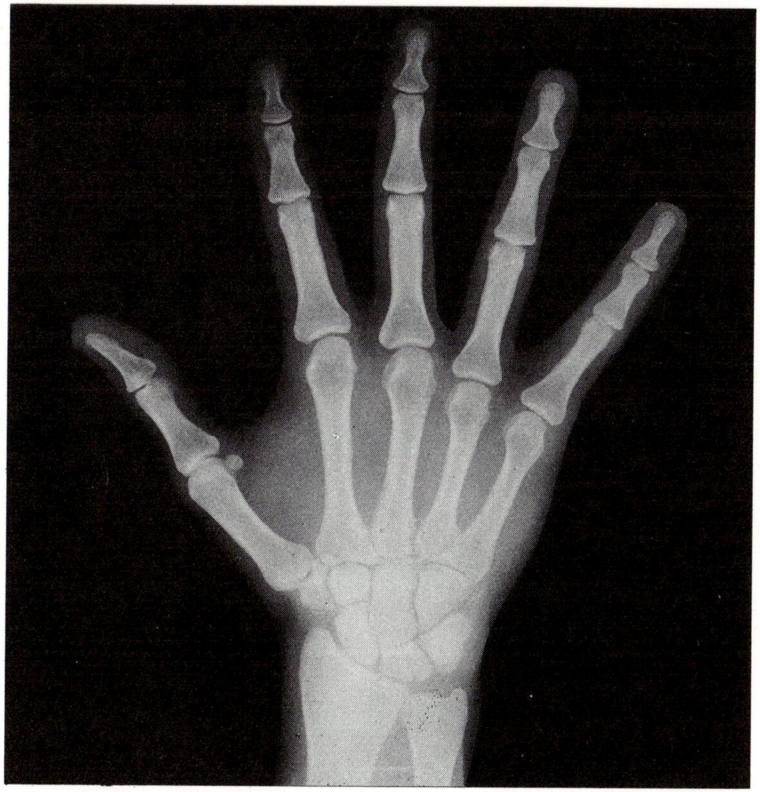

FIGURE 7·20b

FIGURE 7.20b, **the whole arm in** FIGURE 7.18 **and the drawing of the articulation of radius and ulna in** FIGURE 7.21, **describe how you think that the radius and ulna behave when you turn your hand over through 180°.**

D **Give another example of a rotational joint in the skeleton.**

124

FIGURES 7.20a and b are X-ray photographs of two hands, taken from above. The hands were held with the palms flat against an X-ray plate and X-rays were shone through their upper sides. The clearly visible structures are bones; the shadowy parts are flesh and cartilage.

A **Say, giving two reasons, whether they are right hands or left hands (one reason comes from looking at a top view of your own hands).**

B **Leaving aside differences of individual shape, list as many differences as you can see between the phalanges in the two hands.**

C **Apart from differences of individual shape, list as many differences as you can between the wrist bones in** a **and** b.

D **Is it possible to decide from these photographs and the amount of information given about them in this question** alone **whether either or both of these hands are human or not?**

E **Suppose you are told that both photographs** are **of human hands.**

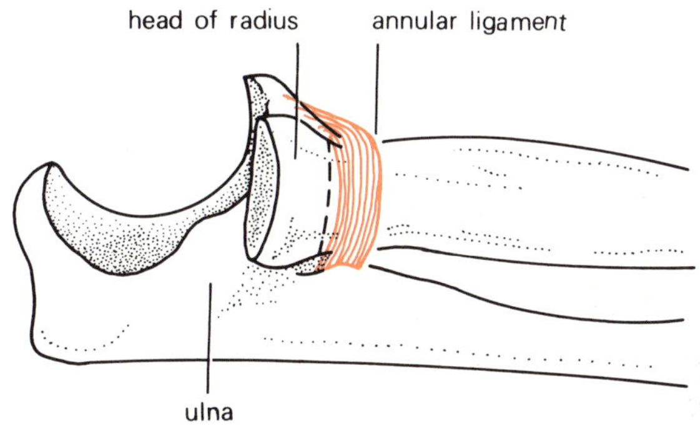

FIGURE 7·21

i **State as many reasons as you can in support of the view that** a **is a younger hand than** b.

ii **State why, from the evidence of the pictures alone, the reasons you give in** i **cannot be more than guesses.**

125

'Biceps' means 'two heads'. This refers to the fact that the biceps muscle has two tendons at the shoulder end, both attached to the shoulder blade.

'Triceps' means 'three heads'. This muscle has three tendons at its upper end, two attached to the shaft of the humerus, the third to the shoulder blades. (FIGURE 7.22—only one tendon is shown)

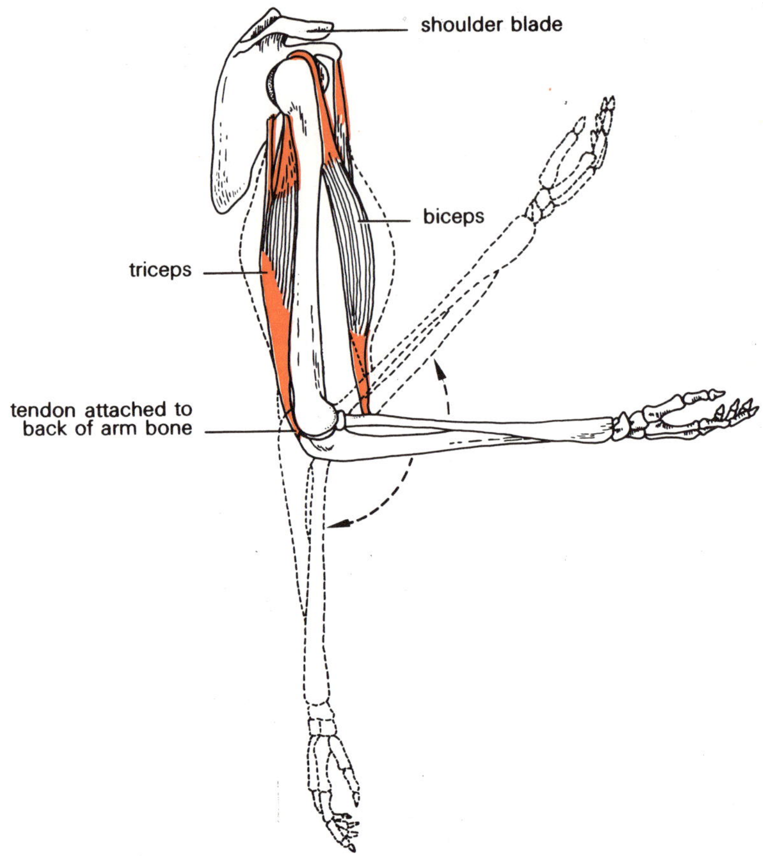

FIGURE 7·22

A Each of these muscles has one tendon at its lower (elbow) end. What are these tendons attached to?

B i When the triceps muscle contracts, it pulls all the tendons which attach it to bones. Explain why this contraction causes the forearm to be lowered.

ii What effect does this movement have on the biceps muscle?

iii What can be said about nervous impulses passing to the biceps at the time when the triceps is contracting?

iv When the triceps has contracted fully, the arm is extended, as in FIGURE 7.18a. **Describe as exactly as you can how the contracted triceps gets back again to its fully stretched condition (as it would be in** FIGURE 7.18c **).**

C FIGURE 7.10c **shows another antagonistic group of muscles.**

i Which are they?

ii What effect on movement at the knee will each muscle of the group have when it contracts?

126

If an isolated muscle is held firmly by its tendon at one end and is left free to move at the other, then when it is stimulated via its nerve, it will contract.

If the isolated muscle is held firmly by its tendons at *both* ends and is stimulated via its nerve it *cannot* contract, but it will develop tension, that is it will pull on both tendons without producing any shortening. (Tension is force per unit cross sectional area of muscle.)

A Imagine you are lifting a heavy suitcase off the ground with your arm fully stretched. Will the nerve impulses to the muscles attached to the shoulder blade and to the arm muscles be causing mainly contraction of the muscle cells or the development of tension in the muscle cells?

B Now imagine that you raise the suitcase by raising your forearm to the position in FIGURE 7.18b. **What will be the position with regard to contraction and with regard to the development of tension in these muscles:**

i during the rise,

ii when the forearm is raised and steady?

C Although we cannot see it in FIGURES 7.12b **and** 7.13, **there are chemical cross bridges between the actin and the myosin rods of the myofibril. It is thought that when myofibrils contract, these bridges are broken. Energy from ATP causes the actin rods to slide inwards (exactly how this happens is not known), and the cross bridges are re-established in the new position. It is something like a ratchet moving up several notches.**

When a muscle develops tension it does not shorten. Probably all the myofibrils remain the same length, and the tension is in some way associated with the strength of the

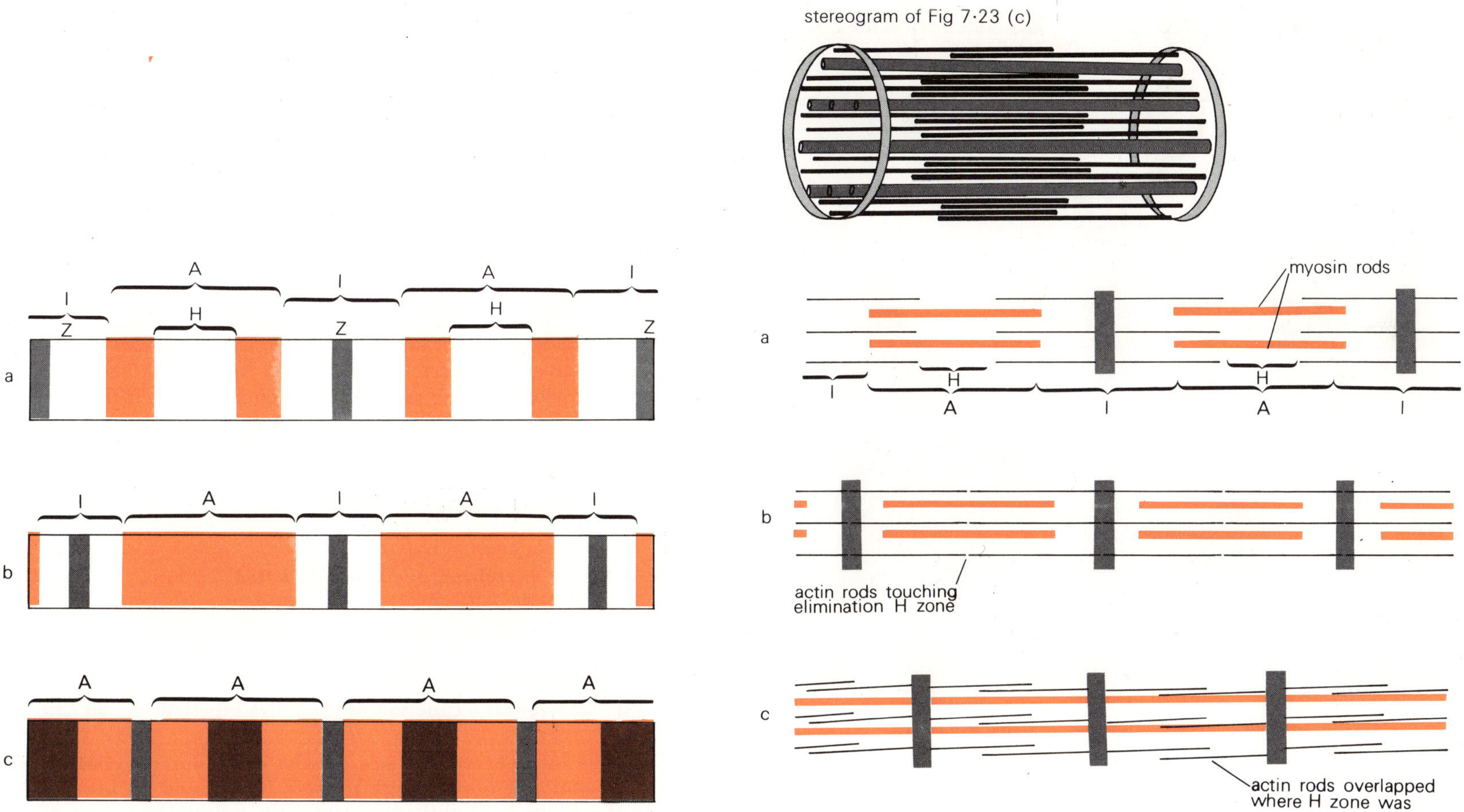

FIGURE 7·23

chemical bridges between the actin and myosin rods.

FIGURE 7.23a shows the relationship of the actin and myosin when the muscle is in the resting state (the length it assumes at rest in the body). The actin rods are some distance from each other, i.e. there is a distinct H band.

In FIGURE 7.23b, the actin rods have drawn in, eliminating the H band, and are just touching each other.

In FIGURE 7.23c we see that the actin rods have drawn in further, eliminating the H band and overlapping each other in the centre of the A band.

If we assume that muscle tension depends on the strength of the chemical cross bridges between actin and myosin rods, try to explain the following two facts:

I Maximum tension in the muscle is reached when the actin rods just touch each other (as in FIGURE 7.23b).

II After the actin rods start to slide over each other in the centre of the A band (as in FIGURE 7.23c), tension falls.

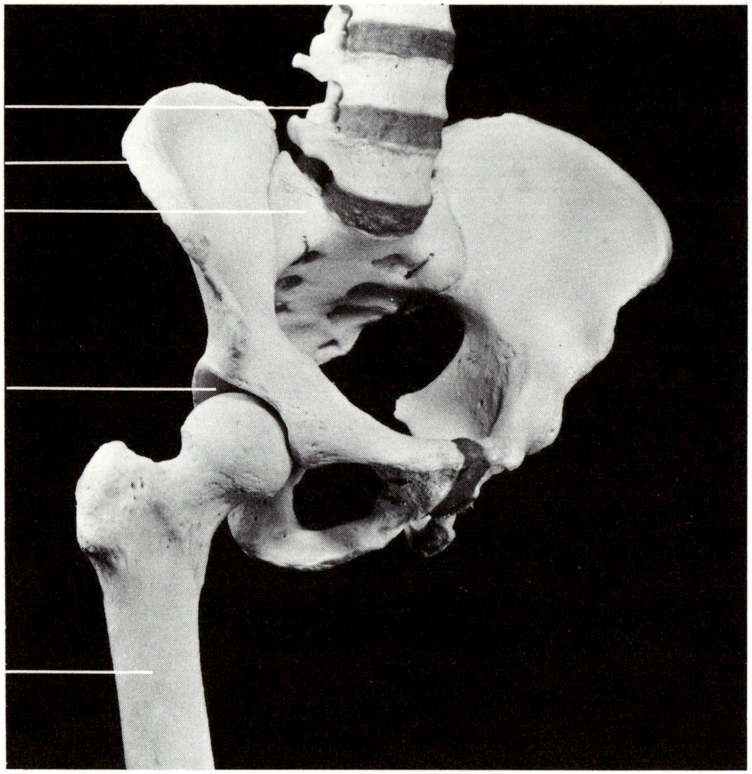

FIGURE 7·24

127

FIGURE 7.24 is a photograph of the human hip bone, the top of the right thigh bone (femur) and the base of the vertebral column. Look back to FIGURE 7.16 of the transparent woman and see where these bones fit in to the complete skeleton.

We are going to compare the structure and function of the shoulder girdle with that of the hip girdle. The back of one half of the shoulder girdle is shown in FIGURE 7.17; the clavicle, which is part of it, cannot be seen here though it can in FIGURE 1.2 (p 10).

A **How does the rest of the skeleton restrict or allow freedom of movement of each of these girdles?**

B **How does the joint of the shoulder blade with the humerus compare with that of the hip girdle with the femur?**

C **Ignore the vertebral column and femur in FIGURE 7.24. Apart from the very different shapes of the complete shoulder girdle and hip girdle, in what way are they strikingly similar?**

D **The main function of the shoulder girdle is to be an anchor for the muscles which move the arm.**

Apart from the joint with the leg, what does FIGURE 7.24 indicate might be the two main functions of the hip girdle?

128

FIGURE 7.25 is a photograph of a human spine from the right side. The first seven vertebrae are the *cervical* vertebrae; the next twelve are the *thoracic* vertebrae; the next five are the *lumbar* vertebrae; the next five *sacral* vertebrae are fused to form the *sacrum* (from the Latin word 'sacer', sacred—the Romans called it this because it was the last part of a cremated body to burn). The last four are the tiny *coccygeal* vertebrae.

Identify the regions in the vertebral column of FIGURE 7.16, the transparent woman (p 108).

A **Which vertebrae are those labelled A, B, C, D, E and F in FIGURE 7.25 ?**

B **i** **From which region of the spine is the close-up shown in FIGURE 7.26 ? State which feature of the vertebrae makes you sure of your answer.**

ii **What is the function of the holes labelled X in FIGURE 7.26 ?**

C **The intervertebral discs consist of an outer (gristle) part, containing many tough protein fibres, and an inner pulp. Each one is slightly compressible, unlike the centra of the vertebrae on either side of them. They constitute about one fifth of the length of the vertebral column, but this is not evenly distributed for they are much thicker in the cervical and lumbar regions than in the thoracic region.**

i **Explain how the presence of the intervertebral discs makes movement of the backbone possible, and say what kinds of movement they make possible.**

ii **How is the fact that the discs and the inner pulp in the cervical and lumbar regions are thick and in the thoracic region are thin related to the amount of movement in these regions?**

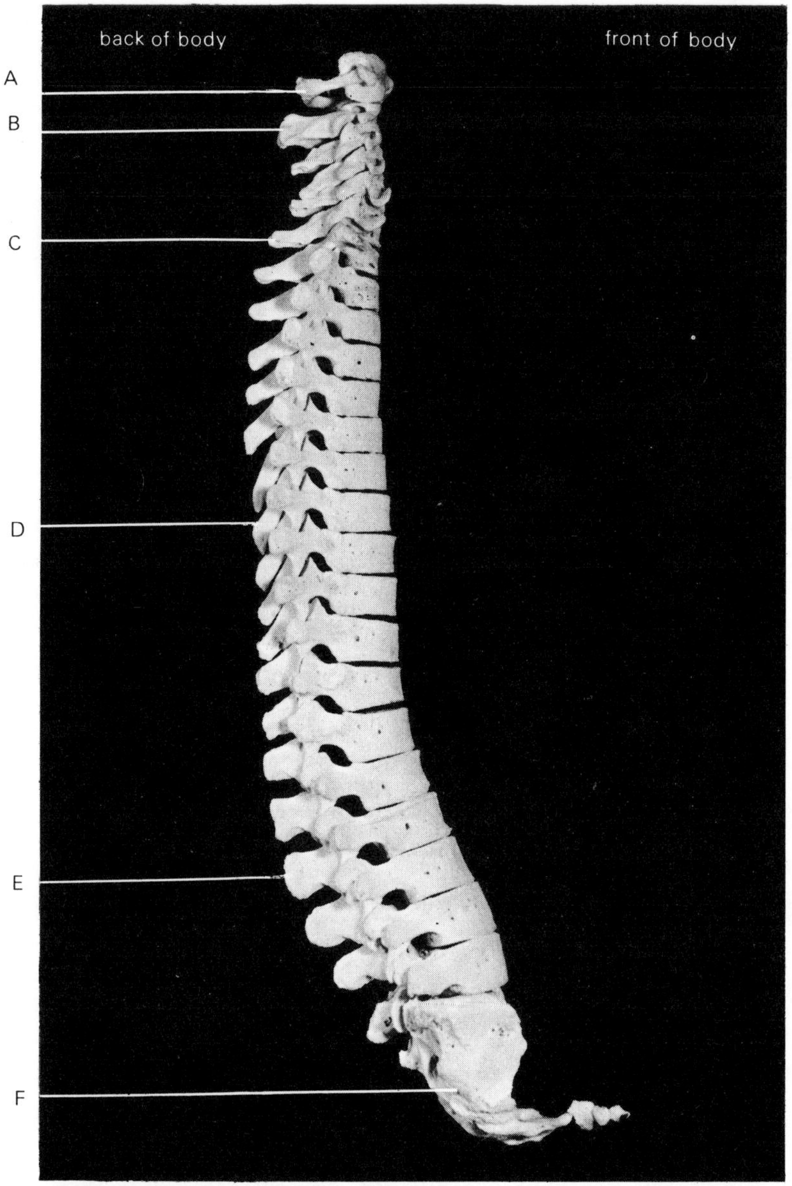

FIGURE 7·25

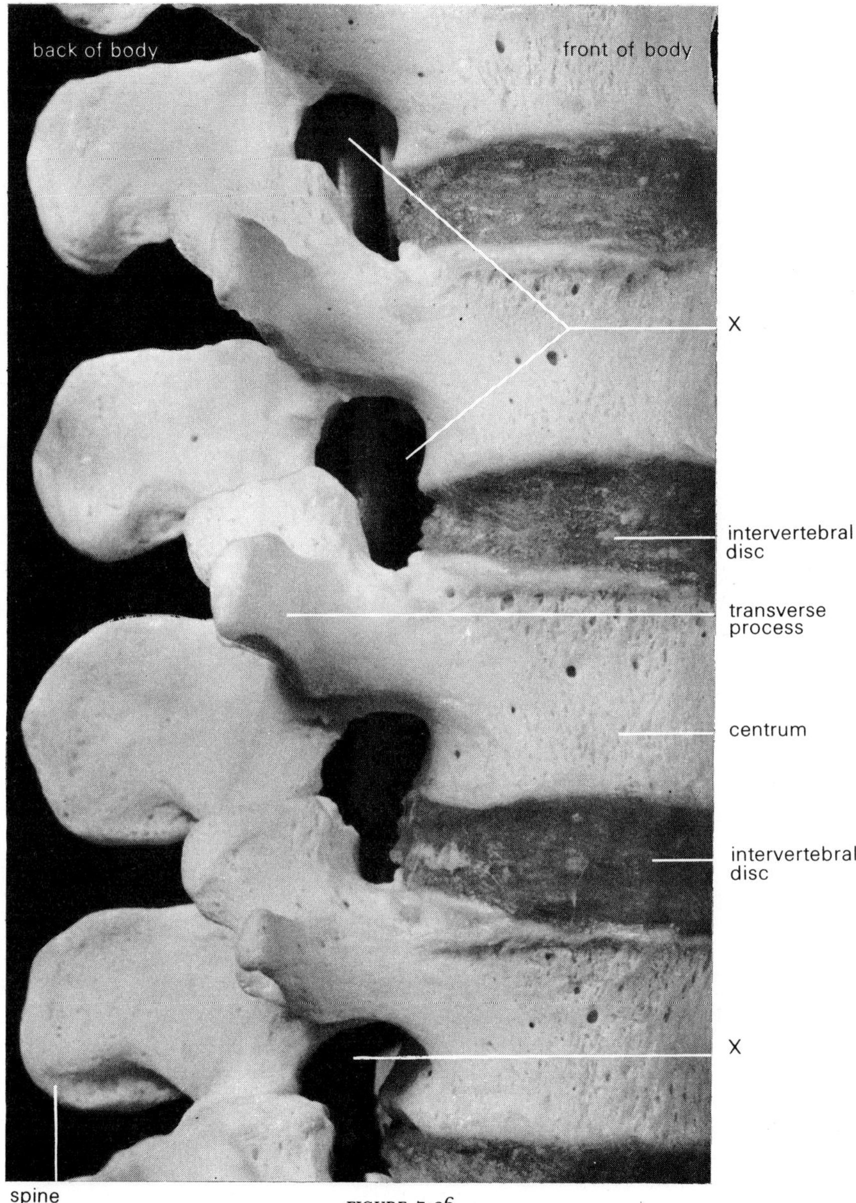

FIGURE 7·26

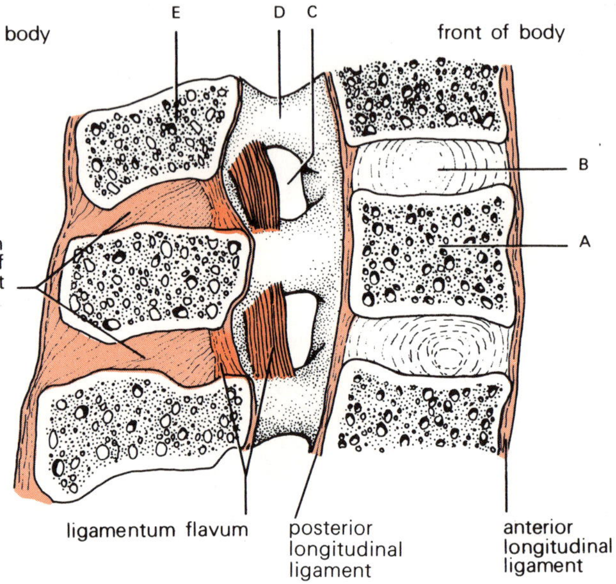

FIGURE 7·27

129

FIGURE 7.27 is a drawing of a longitudinal section through three of the vertebrae seen in FIGURE 7.26. Not only are the bones and intervertebral discs shown, but also the ligaments binding them together.

A **What are the parts labelled** A, B, C, D **and** E **?**

B **Describe the position of the neural canal in** FIGURE 7.27, **stating the structures which form its margins.**

C FIGURE 7.27 **shows that at the front of the backbone there is an** anterior longitudinal ligament. **This is made of very tough protein fibres (like other ligaments and tendons), which are not capable of stretching much. It is bound closely to the intervertebral discs but is only loosely attached to the centra of successive vertebrae.**

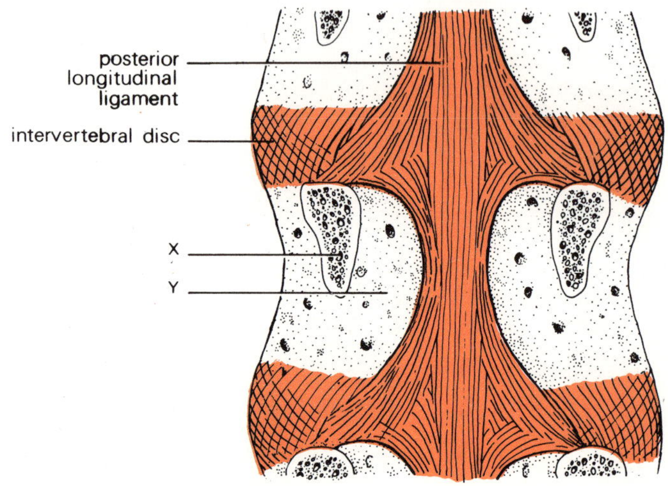

FIGURE 7·28

FIGURE 7.27 **also shows the** posterior longitudinal ligament. **It is similar in shape to the anterior ligament and is shown more fully in** FIGURE 7.28.

i **Find out from** FIGURE 7.27 **what view you are seeing in** FIGURE 7.28. **Then say what structures** X **and** Y **in** FIGURE 7.28 **are.**

ii **From** FIGURES 7.27 **and** 7.28, **describe the appearance of the posterior longitudinal ligament (the anterior longitudinal ligament is similar) and how you think it is connected to the intervertebral discs.**

D **The intervertebral discs adhere to the vertebrae on either side of them, so even if there were no ligaments, the vertebrae would not separate from each other. What functions do you think the anterior and posterior longitudinal ligaments and the ligament surrounding the intervertebral discs have?**

130

A **Find the** ligamenta flava **in** FIGURE 7.27, **and describe their position.**

B **Now look at** FIGURE 7.29, **which shows another view of two of the ligamenta flava.**

i **Say what view of the vertebrae this is.**

ii **Say what the structure** A **is.**

iii **Say how this view supports the description of the position of the ligamenta flava you gave in** A, **and any feature about them which this view adds to that seen in** FIGURE 7.27.

C **The ligamenta flava are made of different material from the other ligaments we have looked at in this book. They too consist of millions of tough protein fibres, but each fibre is elastic, making the whole ligament elastic—stretchable, though resisting stretch; and once stretched, having a strong tendency to return to the former contracted length.**

i **What effect will the ligamenta flava have on the movement of the backbone?**

ii **How do the ligamenta flava help to protect the intervertebral discs?**

iii **How do the ligamenta flava help to keep a man in an upright position?**

D A **Stand with your legs pressed against a wall. Bend forwards as much as you can, keeping your legs pressed against the wall and not bending from the hip. Find out in this way how much your backbone can bend forwards.**

B **Do the same again, this time keeping the fronts of your legs against the wall, and find out how far your backbone can bend backwards.**

C **Stand with your knees together and your legs rigid and bend sideways, thus finding out how much your backbone can bend sideways.**

i **Say what will happen to the anterior and posterior longitudinal ligaments, the ligaments surrounding the intervertebral discs, the discs themselves, and the ligamenta flava when you are in positions** A, B **and** C. **How are these structures helping to limit the extent of movement in each case?**

ii **Look back to** FIGURE 7.25. **What feature of the vertebral column can you see that obviously limits the extent of bending in position** B **?**

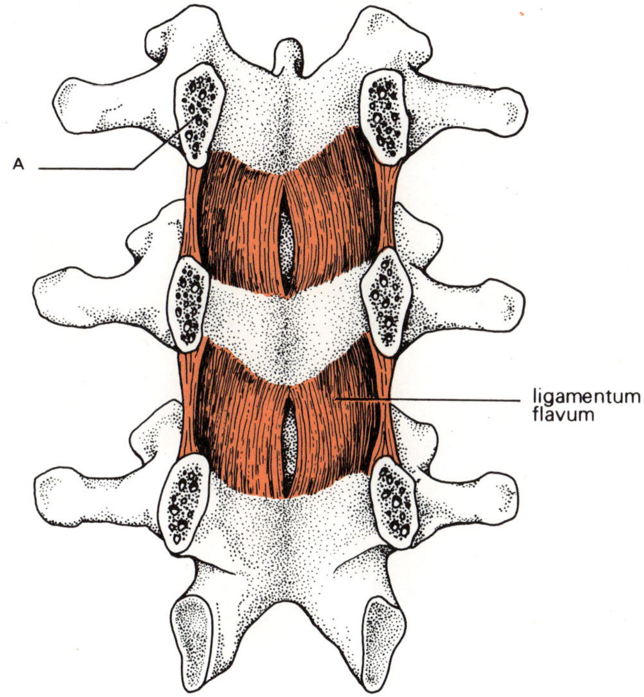

FIGURE 7·29

131

(The following photographs of human vertebrae show holes that were drilled through the centra in order that the vertebral column could be strung together. These should be ignored.)

FIGURE 7.30 shows two lumbar vertebrae. The one on top (a) is an underneath view from the side. The one below (b) is seen flat on from above.

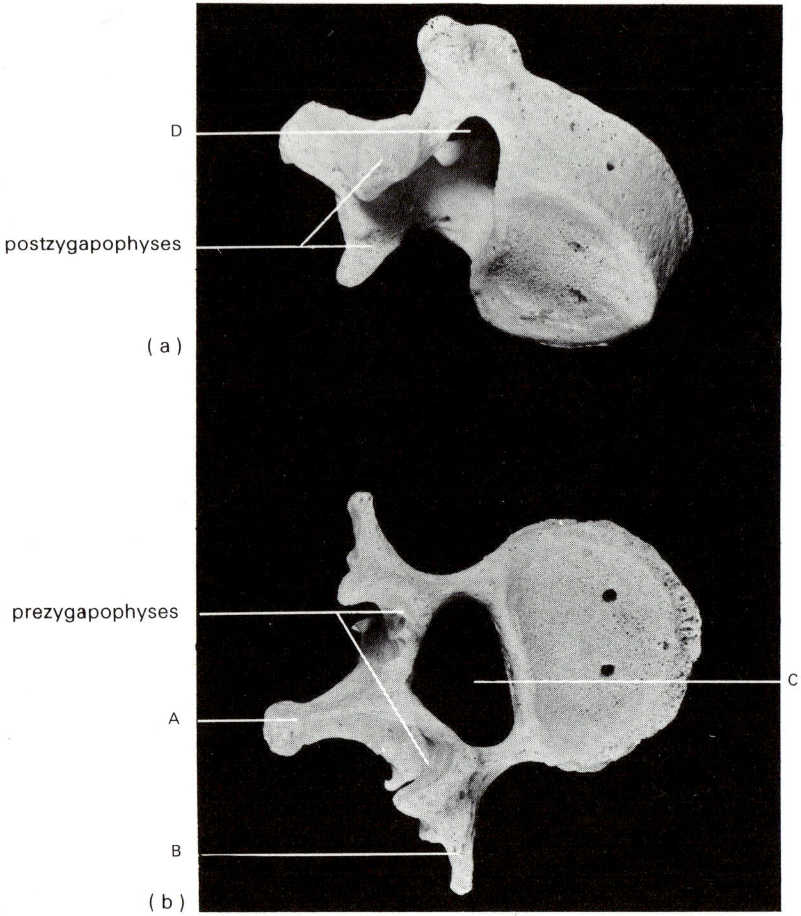

FIGURE 7·30

A **Explain why the views of** a **and** b **are said to be from below and above.**

B i What are the structures in FIGURE 7.30 **labelled** A, B, C, **and** D **?**

ii In life, what structure is situated in the space C **? Why is this a good place for this structure to be situated?**

C **The pre- and post-zygapophyses have been labelled in** FIGURE 7.30. **Write a short description of them. Include in your description the directions in which the surfaces face.**

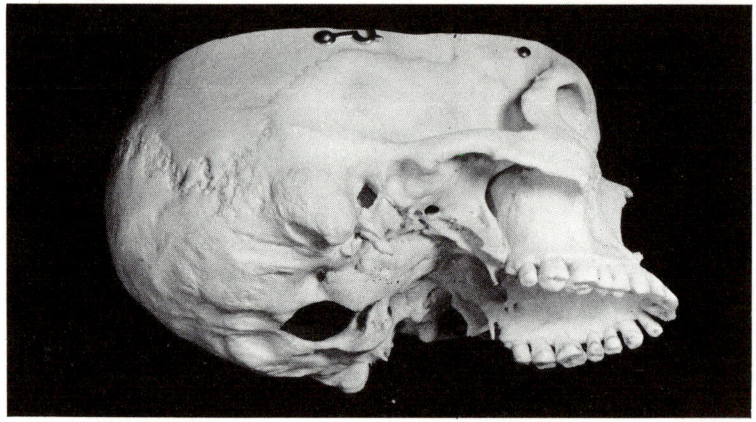

FIGURE 7·31a

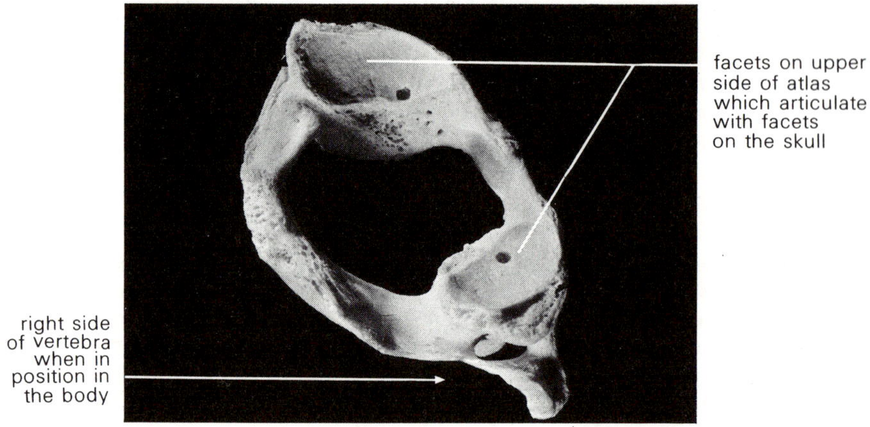

FIGURE 7·31b

132

FIGURE 7.31a is an underneath view of a human skull. FIGURE 7.31b is the top side of the first cervical vertebra (the atlas), photographed from the skeleton's right side (that is, the photograph shows more of the right side of the vertebra than the left).

A i Describe the position and shape of the two facets on the base of the skull which articulate with the top side of the atlas vertebra.

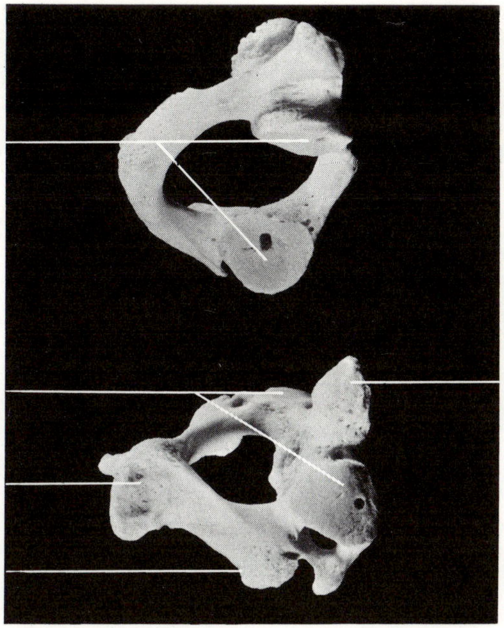

FIGURE 7·32

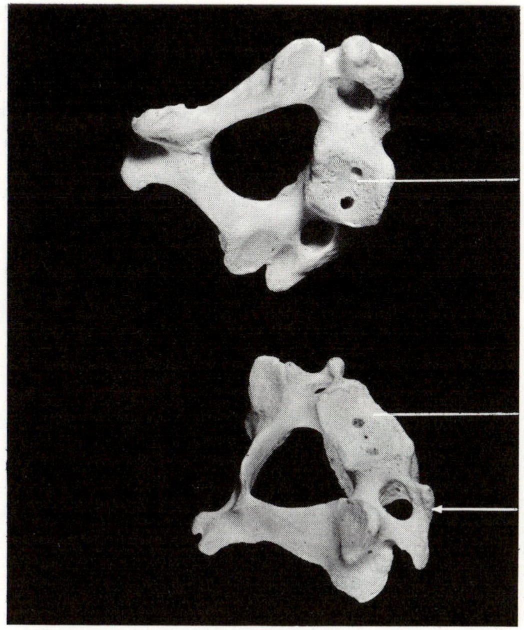

FIGURE 7·33

ii **Say what landmark on the skull makes you sure that your answer to** i **is correct.**

iii **Say what kind of movement this articulation with the skull and the atlas vertebra allows.**

B FIGURE 7.32 **shows in** a **an underneath view of the same atlas vertebra we saw in** FIGURE 7.31b**; and in** b **a view from the skeleton's right side showing most of the features of the upper side of the second cervical (axis) vertebra.**

When you have understood the views from which the photographs have been taken (to help you, refer to FIGURE 7.32 **and try to visualize these two bones in your own backbone), answer the following questions.**

i **Describe the way in which the atlas vertebra articulates with the axis.**

ii **Describe the type of movement that the articulation between the atlas and the axis vertebrae allows.**

iii **State two ways in which the articulation between the atlas and axis vertebrae differs from that which we saw between the lumbar vertebrae in question 131.**

iv **What, then, is the essential function of the odontoid peg, and how does it perform it during life?**

C FIGURE 7.33 **shows in** a **the underside of the axis vertebra whose upper side we saw in** FIGURE 7.32b**; and in** b **a view of the upper side of the next cervical vertebra, taken from the skeleton's right side.**

i **Describe how these two vertebrae articulate with one another.**

ii **Compare the photographs of the atlas vertebra (**FIGURES 7.31b **and** 7.32a**) with those of the lumbar vertebrae (**FIGURES 7.30a **and** b**). State as many** differences **between them as you can discover from these photographs.**

iii **Compare the photographs of the axis vertebra (**FIGURES 7.32b **and** 7.33a**) with those of the lumbar vertebrae (**FIGURES 7.30a **and** b**). State as many** differences **and as many** similarities **between them as you can discover from the photographs.**

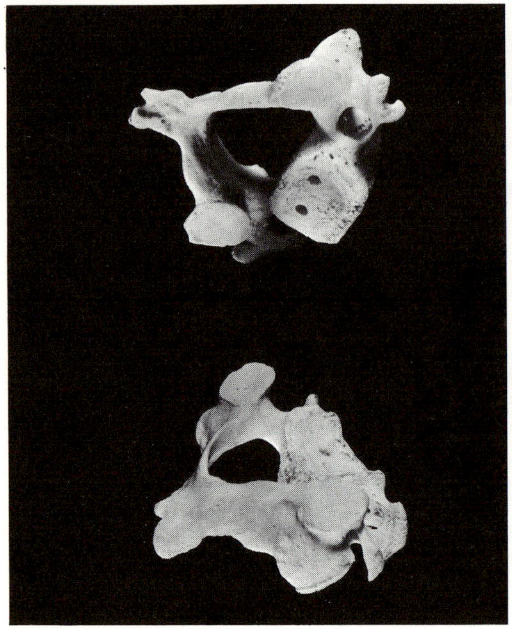

FIGURE 7·34

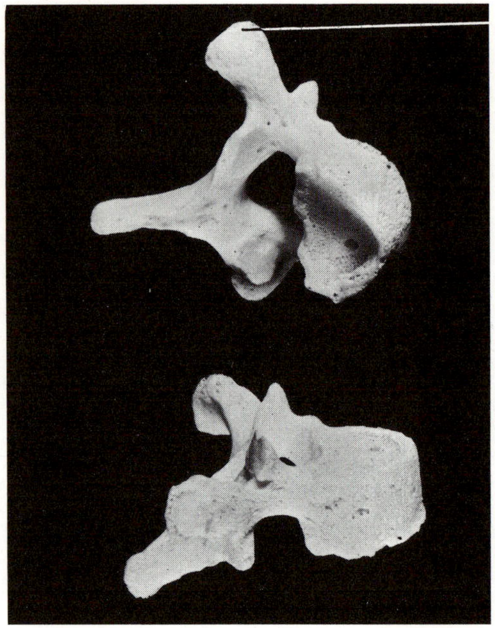

FIGURE 7·35

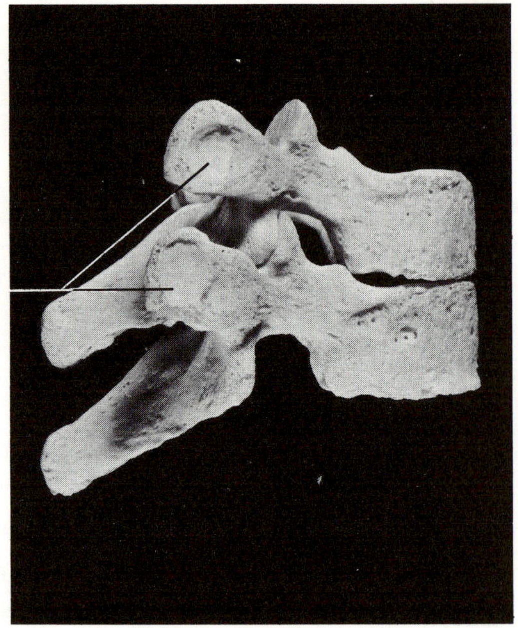

FIGURE 7·36

133

FIGURE 7.34 shows in a the underside of the cervical vertebra whose upper side we saw in FIGURE 7.33b; and in b a view of the upper side of the next (fourth) cervical vertebra, taken from the skeleton's right side.

The atlas and axis vertebrae are specially adapted to enable movements of the skull to take place. The remaining five cervical vertebrae are similar to one another, and FIGURES 7.34a and b give a good idea of what each of them looks like.

A **What feature does each of the cervical vertebrae (including the atlas and axis) possess that is not shared by the lumbar vertebrae? (In fact it is not shared by the thoracic or sacral vertebrae either, as you can see if you look on to** FIGURES 7.35, 7.36 **and** 7.37.)

B **Summarize the position regarding pre- and post-zygapophyses in the atlas, axis and the remaining five cervical vertebrae.**

C **The cervical (neck) region of the backbone has a greater range of movement than the thoracic or lumbar; the sacral region has no range of movement. We have seen in question 128C (p 114) that this is associated in the neck region with thick intervertebral discs containing thick inner pulp.**

i **What types of movement are possible in the neck region? (Try moving yours to find out.)**

ii **Describe the pre- and post-zygapophyses in the cervical vertebrae shown in** FIGURE 7.34 **(include in your description the directions in which the surfaces face).**

iii **Imagine the two cervical vertebrae shown in** FIGURE 7.34 **to be joined by their thick intervertebral disc. Imagine the pre- and post-zygapophyses moving against each other when the neck is moved. How does the shape and the direction in which the surfaces face suit the movements of the neck you described in i ?**

iv **Look at the positions these zygapophyses occupy on the neural arches, and at the neural spines. How do these features permit the neck to be bent backwards and forwards?**

134

A Apply what you have just learned about the zygapophyses and neural spines of the cervical vertebrae to the zygapophyses and neural spines of two adjacent chest (thoracic) vertebrae.

FIGURE 7.35 shows two of these vertebrae separated. As before, a is largely an underneath view; b is almost a side view (showing the right side of the vertebra), though it shows several of the features of the front of the bone.

FIGURE 7.36 shows the two vertebrae close together, as they would be in life (the intervertebral discs in the chest region are thin).

i Describe the shape and position of the pre- and post-zygapophyses of the thoracic vertebrae. Include in your description the directions in which the surfaces face.

ii The thoracic vertebrae permit almost no bending of the chest backwards or forwards.

Remember how in the cervical region the shape and position of the zygapophyses, and their neural spines, did permit this (question 133C, FIGURE 7.34).

How do:

a the shape and position of the zygapophyses and

b the neural spines of the thoracic vertebrae prevent these movements? In answering, look at FIGURES 7.35 and 7.36 and try to imagine movement between these vertebrae.

iii What sort of movement do you think the joint between the zygapophyses of the thoracic vertebrae does allow? Give the reason for your answer.

B What is the structure labelled A in FIGURE 7.35 ?

C The surfaces X in FIGURE 7.36 are smooth and are articulating surfaces. With what do they articulate?

D Compare FIGURE 7.26, the side view of the lumbar vertebrae, with FIGURE 7.36, the side view of the thoracic vertebrae. Disregarding relative size of the whole bone, and disregarding the zygapophyses, what are the two main features of difference between them?

E FIGURE 7.36 does not show the intervertebral disc, which is thin between thoracic vertebrae. But even without it, it is obvious that the position of the joint between the pre- and post-zygapophyses in the thoracic vertebrae is different

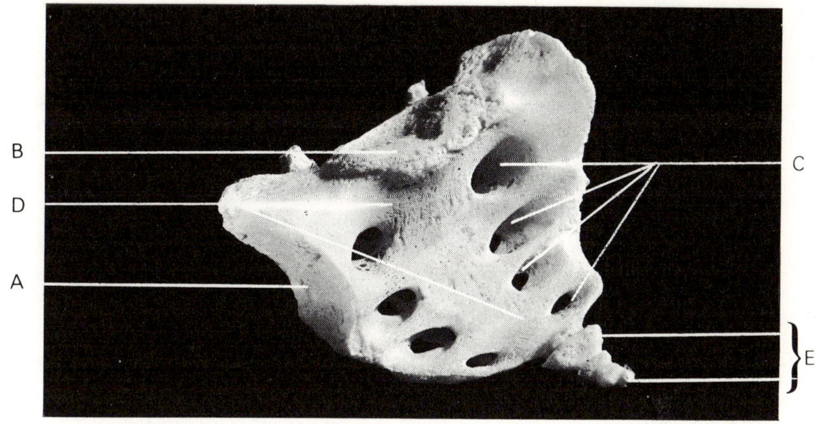

FIGURE 7·37

from the position in the lumbar vertebrae (FIGURE 7.26).

Say:

i what the position of the joint is in each case;

ii how the structure of the lumbar vertebrae positions this joint;

iii how the position of the joint in the case of the lumbar vertebrae permits bending of the backbone forwards and backwards while in the case of the thoracic vertebrae it prevents it.

135

FIGURE 7.37 is a photograph of the front of the sacrum taken from the skeleton's right side. Look back to FIGURES 7.24 and 7.25 (pp 114–115) and find the position of the sacrum.

A From FIGURE 7.24, what is the main function of the sacrum?

B The sacrum is a single structure. During an early stage of development of the embryo, the parts which give rise to vertebrae become connected in this region and develop together, forming the sacrum. It is not five separate bones joined together but one bone produced by the combined activities of five of the vertebra-forming centres of the embryo.

We cannot see the top side of the sacrum in FIGURE 7.37, but there are knobs corresponding to neural spines and to combined pre- and post-zygapophyses.

i **What parts of the five sacral vertebrae are represented by the region** D **in** FIGURE 7.37 **?**

On each side of this region there are four holes (C**). There are similar holes between other vertebrae in the vertebral column, though they are not placed underneath as they are here.**

ii **How do these holes tell us that the sacrum represents** five **sacral vertebrae?**

iii **Look back to** FIGURE 7.26 **(lumbar vertebrae p 115) and** 7.36 **(thoracic vertebrae p 120) and decide what feature there these holes correspond to. What, then, is their function?**

iv **What is the function of the surface** A **in** FIGURE 7.37 **?**

v **What is the structure** E **in** FIGURE 7.37 **?**

We have seen in questions 136 131-135 ways in which different vertebrae differ from each other. Each one of the 33 is unique. Some differ greatly from each other (for example, the atlas and the thoracic, or the lumbar and the sacral), others only slightly (for example the third and fourth lumbar). All have a common ground plan of centrum, neural arch, canal and spine, transverse processes and zygapophyses.

Variation on a ground plan is common in the living world. It is evidence of evolution by descent from a common ancestor. Biologists think in terms of ground plans and the extent to which individuals do or do not fit into them. The plans exist at all levels. For example, it is thought that the genetic information in all living cells is duplicated in a similar way; different animal and plant cells obviously have a common ground plan; and the vast range of insects all conform to a common basic plan of body organization.

The limbs of amphibians, reptiles, birds and mammals all have a common ground plan, called the *pentadactyl limb*. In this series of questions we are going to look at some of these limbs.

FIGURE 7.38 shows a front view of a human leg. The regions A, B, C and D are four components of the pentadactyl limb. When looking at other pentadactyl limbs, we shall try to find these regions, however much they may have been modified.

A **Write a short description of each of the regions** A, B, C

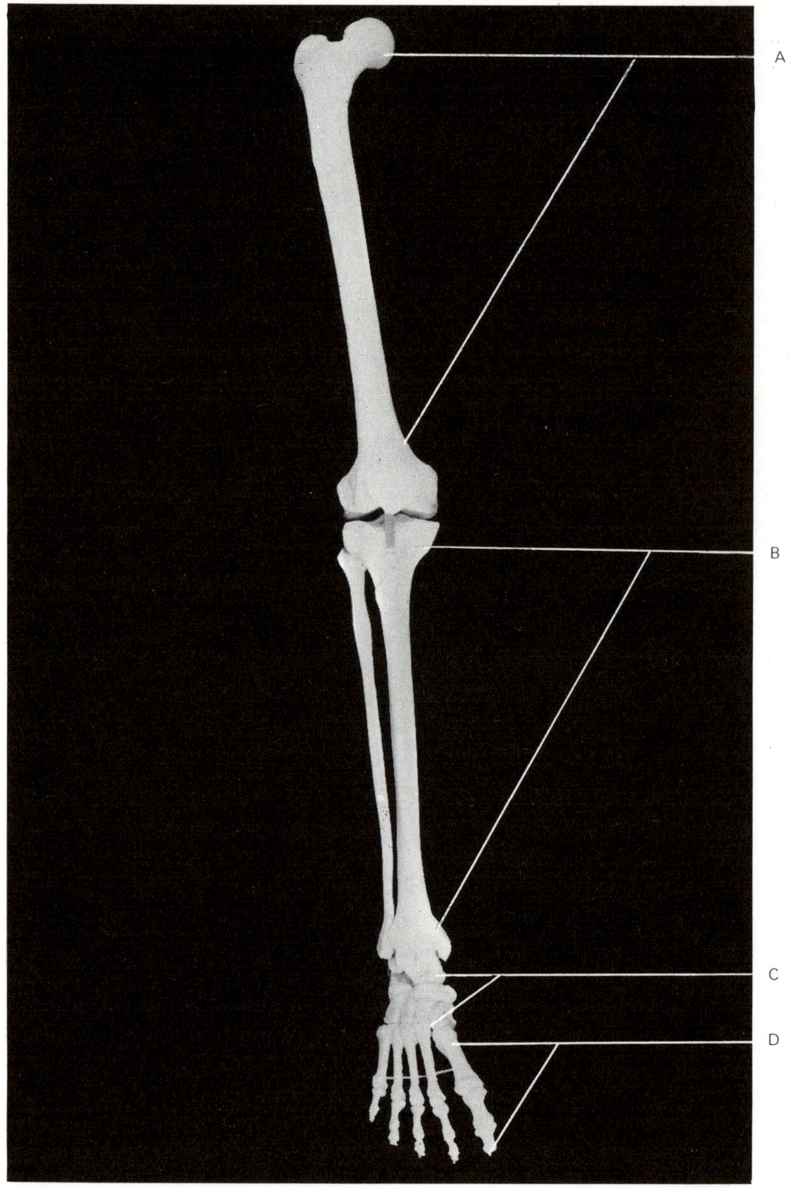

FIGURE 7.38

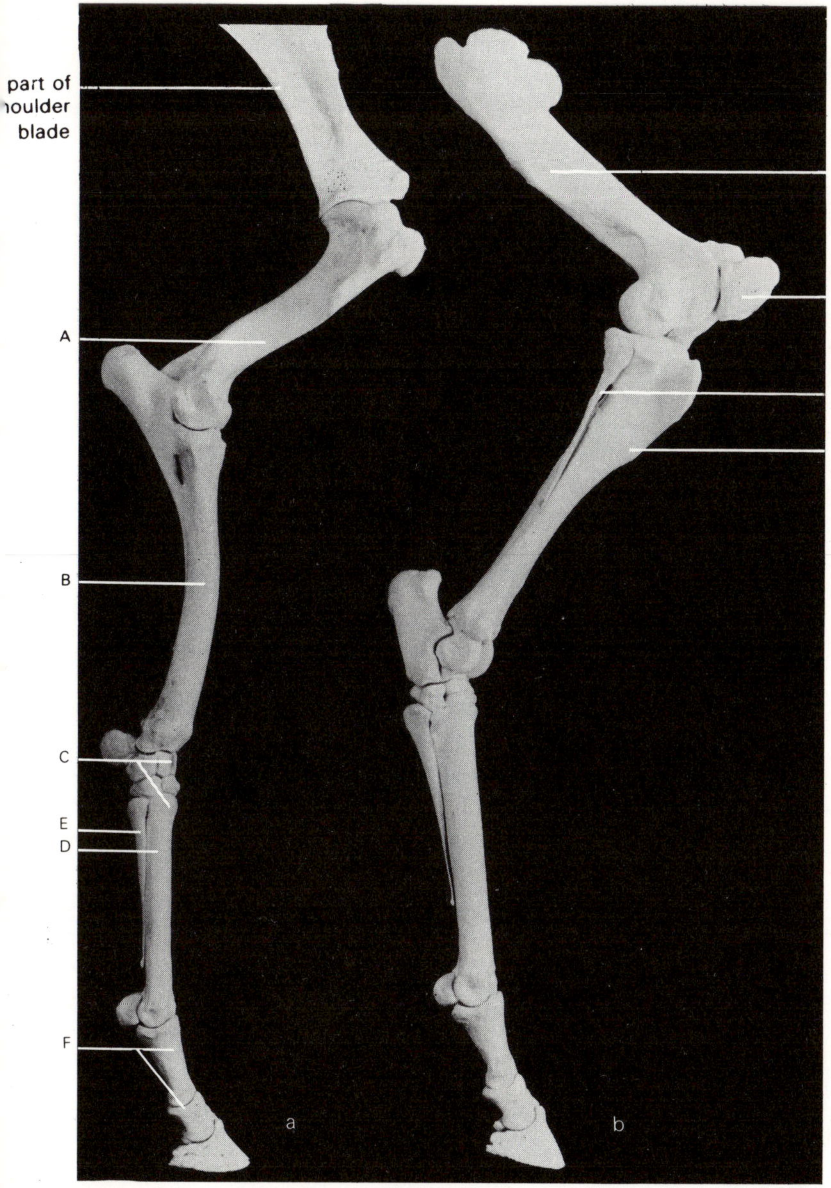

FIGURE 7·39a & b

and D **in** FIGURE 7.38 **and say how the corresponding regions in** FIGURE 7.18 **(of the human arm) compare with them.**

B **Place one hand flat down on your desk. Note the position of your thumb.**

i **How does its position differ from that of your big toe when your foot is flat against the ground?**

ii **How is the range of movement of the thumb and big toe related to the tasks they are called on to perform?**

137

FIGURE 7.39a shows a side view of the fore limb of a horse together with part of the shoulder blade. FIGURE 7.39b shows a side view of the hind limb and does not include the hip girdle.

A **In** FIGURE 7.39a **the shoulder blade has been labelled. The hoof has a similar origin to the nail of the human finger, What, then, are the parts labelled** A–F **?**

B **Apart from proportions, how are the humerus and the radius and ulna**

i **like,**

ii **unlike the corresponding human bones** (FIGURE 7.18a)**?**

C **How is the difference in the structure of the shoulder joint of the horse and the human related to the kind of movement which the humerus makes in each case?**

D **The part of this limb which shows most modification on the pentadactyl pattern is the 'hand'. How is this**

i **similar to,**

ii **different from the human hand?**

E **What are the bones labelled** A, B, C **and** D **in** FIGURE 7.39b **?**

F **What feature of** FIGURE 7.39b **makes it clear that this is a right leg, not a left leg?**

G **The horse fore and hind limbs are very like each other, more so than those of the human. This applies particularly to the bones below the radio-ulna and the tibia-fibula. But how does the arrangement of the humerus and the radio-ulna** differ **from that of the femur and the tibia-fibula?**

138

Our planet has existed for about 4,500 million years. It is not known how long life has been present on it. About 600 million years ago there was a vast increase in the number and kind of marine life forms (in the Cambrian period, see FIGURE 7.40a). Nearly all the present day invertebrate phyla have been found as fossils in the Cambrian. About 500 million years ago, in the Ordovician period, the first vertebrates, fish, appeared. These were followed by the amphibia (in the Devonian, 400 million years ago), the reptiles (in the Carboniferous, 350 million years ago) and the mammals (in the Cretaceous, 135 million years ago). Check this timetable on FIGURE 7.40a.

For the long Mesozoic era (about 160 million years), the reptiles were the dominant vertebrates. They succeeded in the air and the sea as well as on the land and produced both enormous and bizarre forms. The Mesozoic is called the Age of Reptiles. For quite a long part of the Mesozoic, from the Jurassic onwards, mammals existed side by side with the reptiles, but they were small (the largest being about the size of a rat), nocturnal, probably shy creatures living on insects. Then suddenly at the close of the Mesozoic, most of the reptilian orders, including all the aquatic, aerial and giant land orders, died out. The cause of this is unknown and is one of the greatest puzzles in the geological history of the earth. But their disappearance left the way open for the sudden and diverse radiation of the mammals (FIGURE 7.40b). Within the short span of about 10 million years most of the present day orders of placental mammals had appeared.

A **From** FIGURE 7.40b, **which of the present day mammalian orders arose after the Palaeocene epoch?**

B **From** FIGURE 7.40b, **describe the geological history of the mammalian order Condylarthra, i.e. when it arose, when it flourished, when it became extinct.**

C **Which of the orders shown arose from the condylarth stock?**

139

FIGURE 7.41a is a fossil skeleton of a condylarth called *Phenacodus*. FIGURE 7.41b is a closeup of its hind limbs and FIGURE 7.41c is a reconstruction of the animal. It was about 1·5 m long and had similar dimensions to a modern tapir.

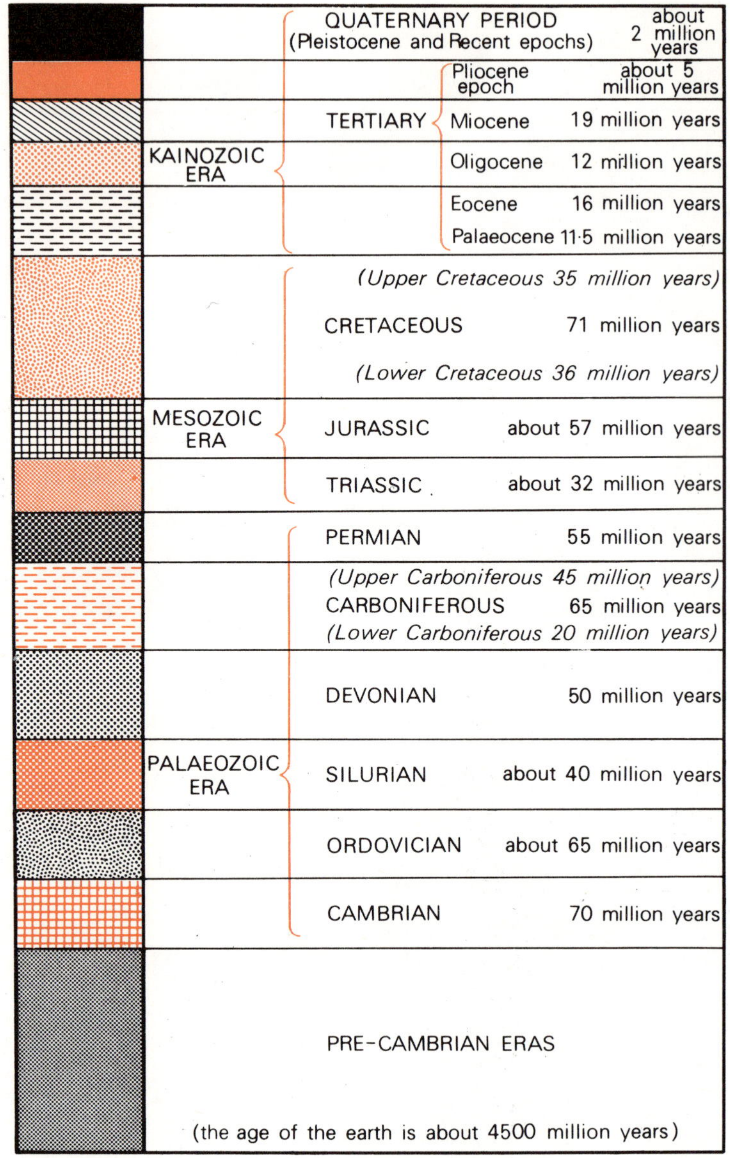

FIGURE 7·40a

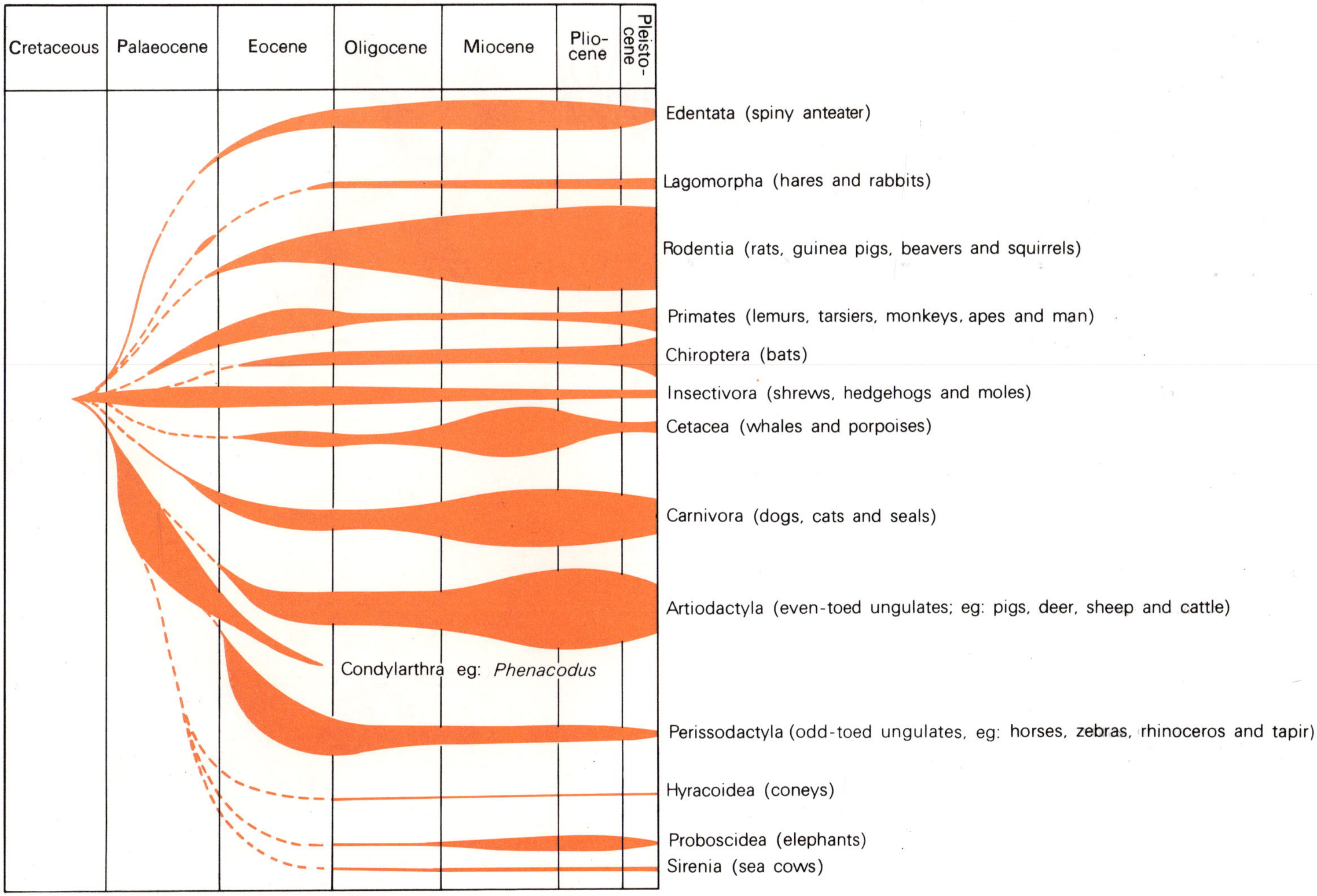

Cretaceous
Palaeocene
Eocene
Oligocene
Miocene
Plio-cene
Pleisto-cene
Edentata (spiny anteater)
Lagomorpha (hares and rabbits)
Rodentia (rats, guinea pigs, beavers and squirrels)
Primates (lemurs, tarsiers, monkeys, apes and man)
Chiroptera (bats)
Insectivora (shrews, hedgehogs and moles)
Cetacea (whales and porpoises)
Carnivora (dogs, cats and seals)
Artiodactyla (even-toed ungulates; eg: pigs, deer, sheep and cattle)
Condylarthra eg: *Phenacodus*
Perissodactyla (odd-toed ungulates, eg: horses, zebras, rhinoceros and tapir)
Hyracoidea (coneys)
Proboscidea (elephants)
Sirenia (sea cows)

FIGURE 7·40b

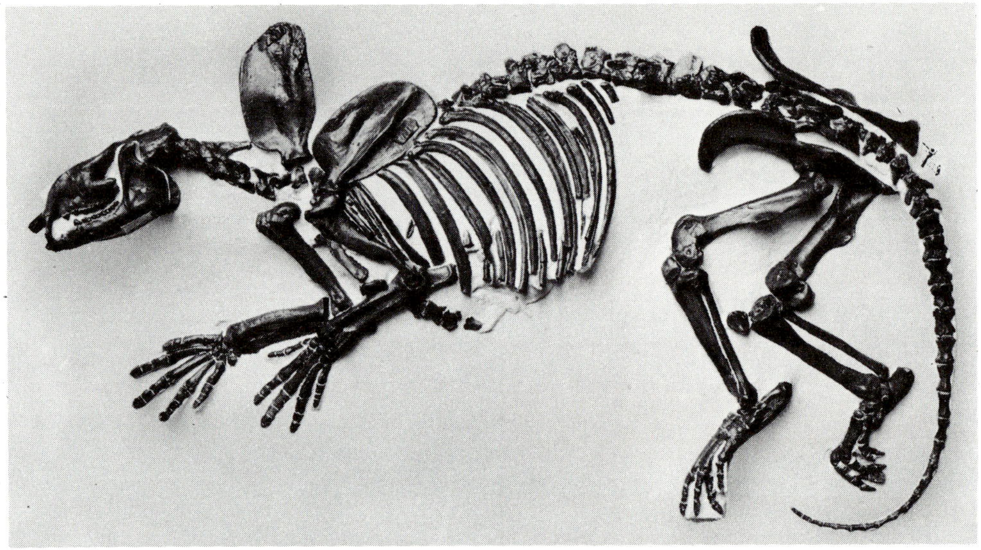

FIGURE 7·41a

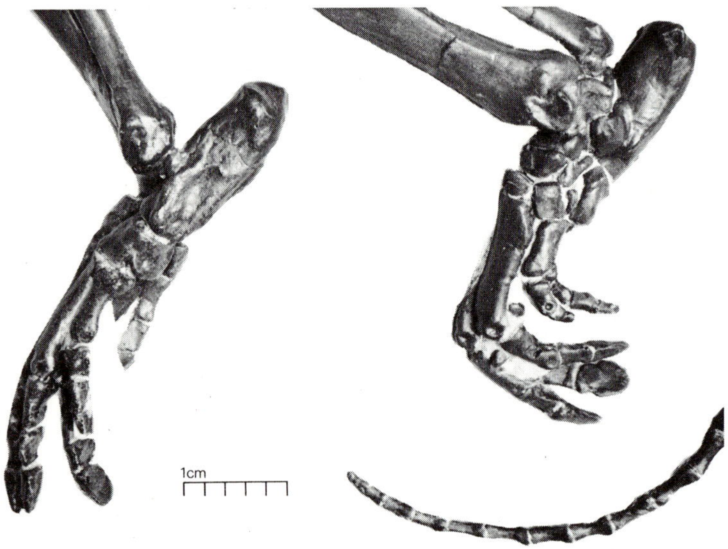

FIGURE 7·41b

FIGURE 7·41c

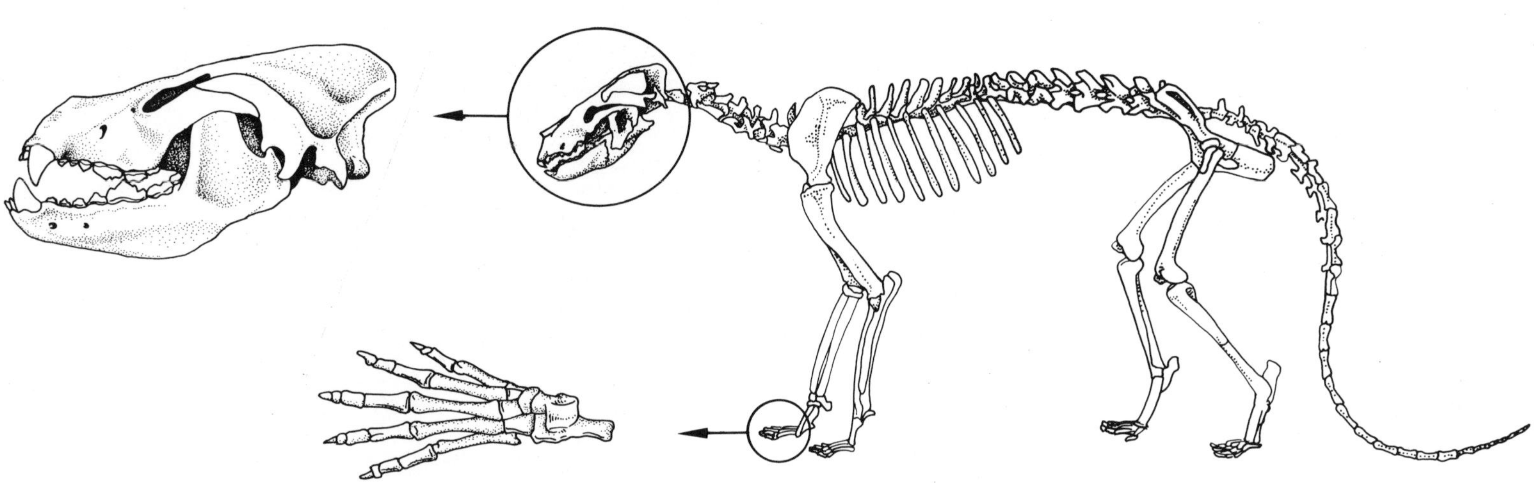

FIGURE 7·42a

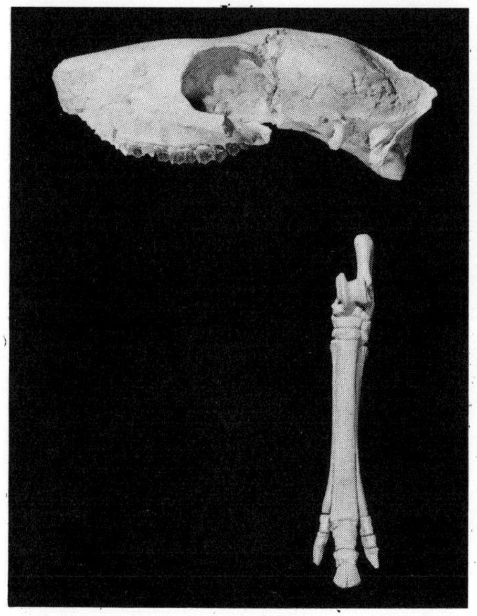

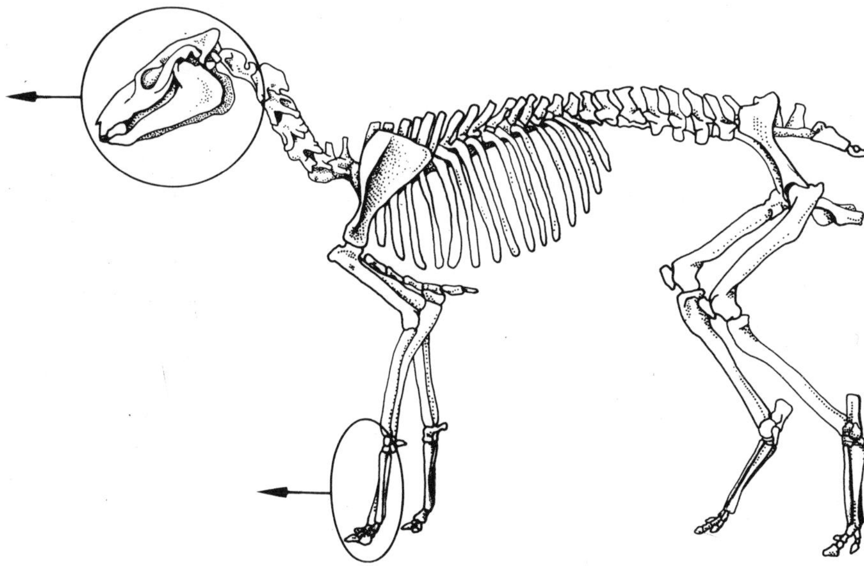

FIGURE 7·42b

FIGURE 7.40b shows that the condylarth and carnivore stocks were separate, though both had a common origin from the primitive mammals of the Cretaceous. In the early days of the study of fossil mammals (at the beginning of this century), this distinction was not obvious, and *Phenacodus* was at various times classified with the primates, the insectivores and the carnivores.

A **How does** FIGURE 7.40b **show that** Phenacodus **would be likely to be confused with the primates, insectivores and carnivores rather than with the other mammalian orders shown there (viz the Edentata, Lagomorpha, Rodentia, Chiroptera, Cetacea, Artiodactyla, Perissodactyla, Hyracoidea, Proboscidea and Sirenia)?**

B FIGURE 7.42a **shows the skeleton of a Miocene bear-dog called** Daphoenodon. FIGURE 7.42b **shows the skeleton of an Oligocene horse called** Mesohippus. **Compare the skeleton of** Phenacodus (FIGURES 7.41a **and** b) **with those of these two animals (one a carnivore, one a herbivore) and say in what ways** Phenacodus

i resembled

ii differed from each regarding general body proportions, tail length, dentition (as far as can be seen) and structure of the foot.

140

Phenacodus is the best-known condylarth. FIGURE 7.40b shows that from this stock arose the Perissodactyla, the present day horses, tapirs and rhinoceroses. The fossil history of horses is better known than that of any other mammalian group. While the skeleton of *Phenacodus* is very different from that of a present day horse, a series of animals is known between the Palaeocene and the present day which show progressively greater specialization away from the *Phenacodus*-type skeleton and towards the horse-type skeleton.

The horse is adapted for rapid movement on hard ground. It browses on grass—it is naturally a prairie dweller—and its teeth and gut are adapted for dealing with this difficult food. It cannot fight and evades its predators by outrunning them.

Compared with a modern horse, *Phenacodus* was a heavy, lumbering animal with stumpy legs (FIGURE 7.41c).

FIGURE 7.42a showed that *Daphoenodon* walked on the digits of its feet which, like those of a present day cat, were held flat against the ground.

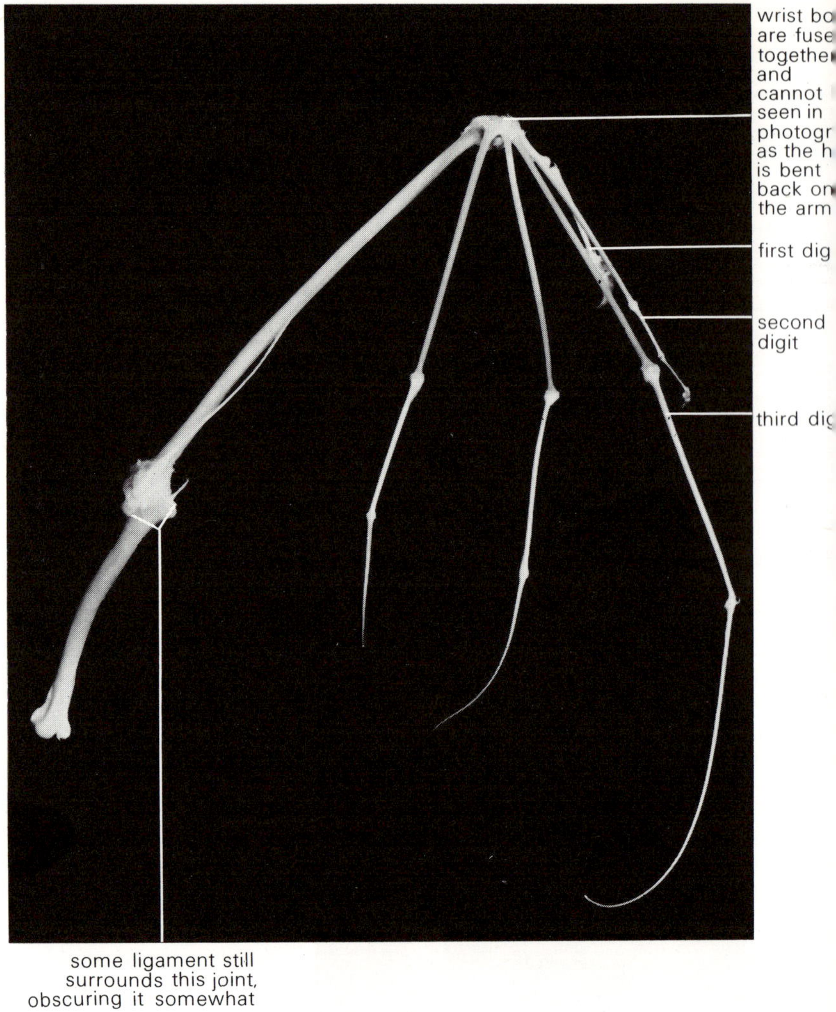

FIGURE 7·43a

A **From** FIGURE 7.41c, **which part of its foot did** Phenacodus **walk on?**

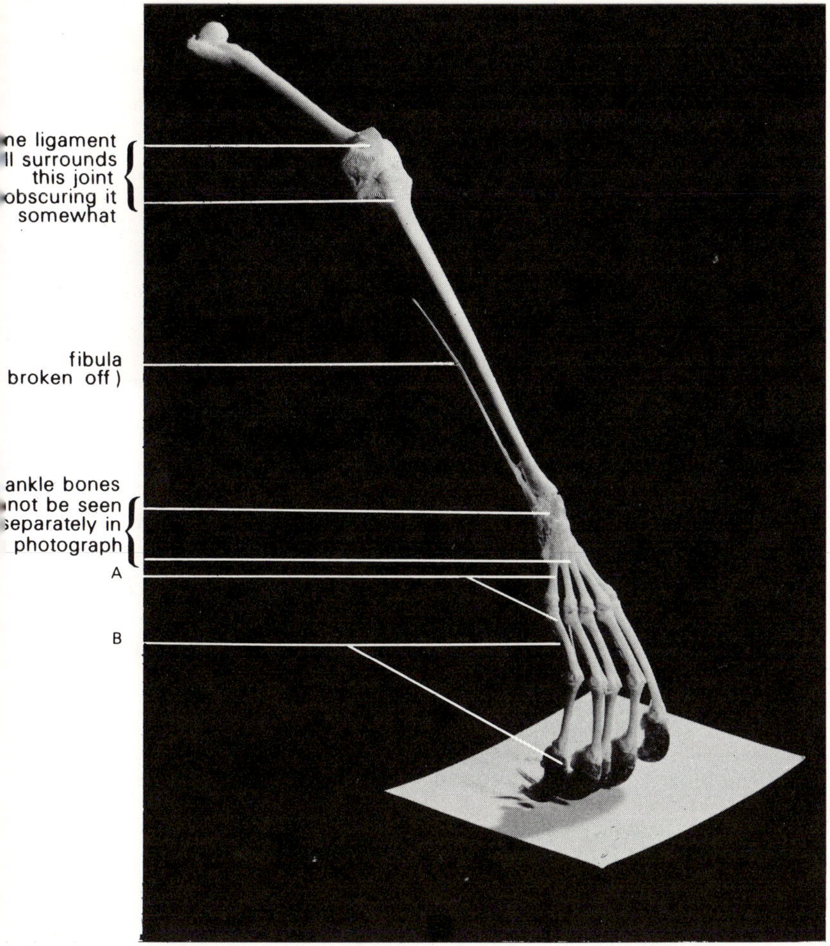

FIGURE 7·43b

B **From** FIGURES 7.39a **and** b **(p 123), which part of its foot does the present day horse walk on?**

C **How is the difference noted in questions A and B related to the fact that the horse is adapted to movement on hard ground while** Phenacodus **was more suited to movement on soft ground?**

D **Assuming that** Phenacodus **was typical of the condylarth stock from which horses have evolved, what changes have taken place in the course of this evolution with regard to:**

- **i** **relative proportions of humerus to radio-ulna and of femur to tibia-fibula**
- **ii** **the relationship of the radius to the ulna and of the fibula to the tibia**
- **iii** **the number of metacarpals and metatarsals?**

E **The horse limbs are long compared with those of** Phenacodus. **This means that the animal can take long strides and so move fast. Comparing** FIGURE 7.41a **with** FIGURES 7.39a **and** b, **which parts of the horse limbs have become elongated?**

141

FIGURES 7.43a and b are photographs of the skeleton of a fore and of a hind limb of a bat. FIGURE 3.9a is a drawing of a bat hanging with its wings extended.

A **The wing is a fold of skin. From** FIGURE 3.9a **(p 47) describe how the wing is supported.**

B **i** **What does** FIGURE 7.43a **show about the ways in which the skeleton of the fore limb conforms to the pentadactyl pattern?**

- **ii** **In what ways is the skeleton of the forelimb specially modified in connection with flight?**

C **Look at the skeleton of the hind limb** (FIGURE 7.43b). **What are the bones labelled** A **and** B **?**

D **Each of the toes ends in a claw. Bats are the only mammals that live in the air. Unlike birds, they are not masters of two kinds of movement. They are much more confined to the air than birds and seldom touch the ground.**

- **i** **In what way does the hooked hind limb assist the bat in its life in the air?**
- **ii** **In what ways do the wings and the hind limbs prevent activity on the ground?**

E **The first digit of the hand is free in all bats; it does not help to support the wing, and is clawed. In fruit bats, the second digit is also clawed. What functions do you think these claws might have:**

- **i** **in movement,**
- **ii** **in defaecation when the bat is at rest?**

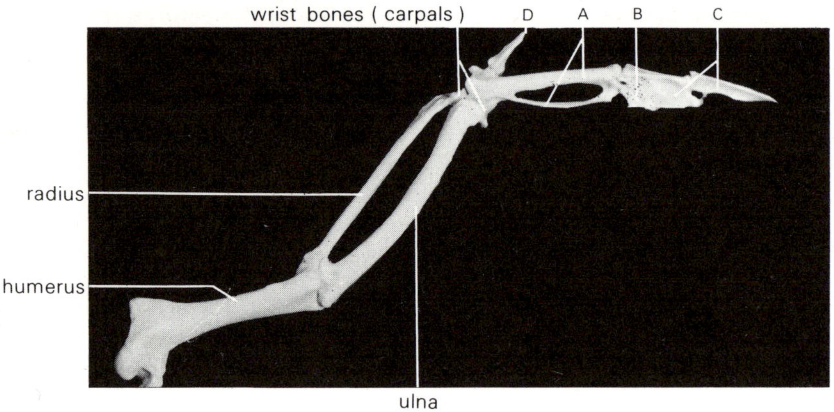

FIGURE 7·44a

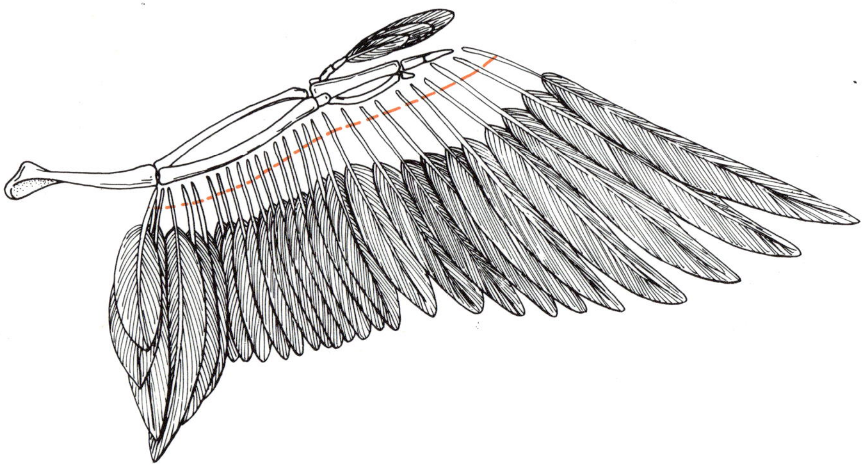

FIGURE 7·44b

142

FIGURE 7.44a is a photograph of the skeleton of the right wing of a pigeon. The trailing edge is that nearer the bottom of the page. Feathers are attached along the edge of the ulna and the bones beyond it, as shown in FIGURE 7.44b.

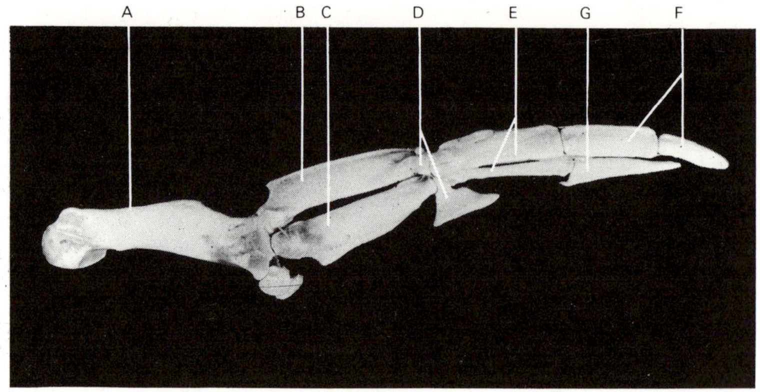

FIGURE 7·45

A **Which parts of the skeleton of this limb obviously conform to the pentadactyl pattern?**

B **How do the wrist bones compare with those we saw in the human skeleton?**

C **From their positions, which parts of the pentadactyl limb must bones A, B and C be?**

D **Name the parts of the skeleton of the fore limb that support the feathers of the wing.**

143

FIGURE 7.45 is another skeleton of the fore limb of a bird.

A **Name the parts labelled A–F.**

B **What is striking about the shapes of the bones in this wing compared with those of the pigeon wing?**

C **Consider the following facts about the skeleton of this wing and then suggest**

i **what you think the limb is used for,**

ii **which bird it belongs to.**

I **The bones are all united. The only movement that the limb makes is at the shoulder joint.**

II **The limb bones of most birds are hollow, and air-filled, making them light. The bones of this wing are solid.**

8 Kidney and reproductive system

144

FIGURE 8.1 is a longitudinal section through the middle of a man's body. It shows some of the excretory and reproductive organs.

A **What are the structures labelled** A, B **and** C **?**

B **Look back to** FIGURE 7.24 **(p 114).**

i **What structures in** FIGURE 7.24 **can be seen in** FIGURE 8.1 **?**

ii **Describe how the following structures lie relative to the bony hip basin seen in** FIGURE 7.24**: the spongy tissue of the penis; the vas deferens; the bladder; the rectum.**

C FIGURE 8.2 **shows a similar section through the body of a woman.**

i **What are the structures labelled** A, B, C **and** D **?**

ii **State which structures, from the front of the body to the back, emerge through the hip basin.**

iii **The vagina is the birth canal. The baby passes through it head first on its journey into the outside world. For this to happen, the vagina must stretch a great deal. But the skeleton cannot stretch. From** FIGURE 7.24, **describe the bony hole through which the baby must pass during childbirth.**

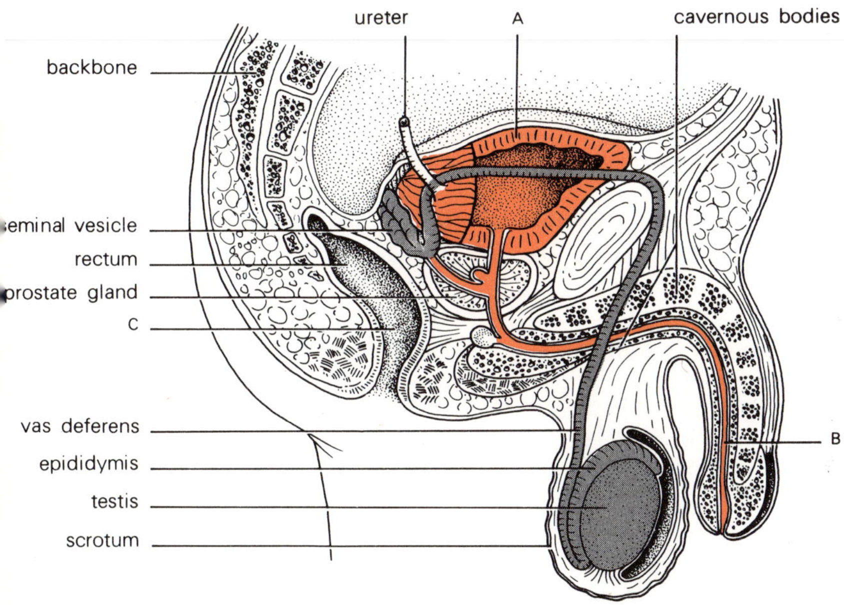

FIGURE 8·1

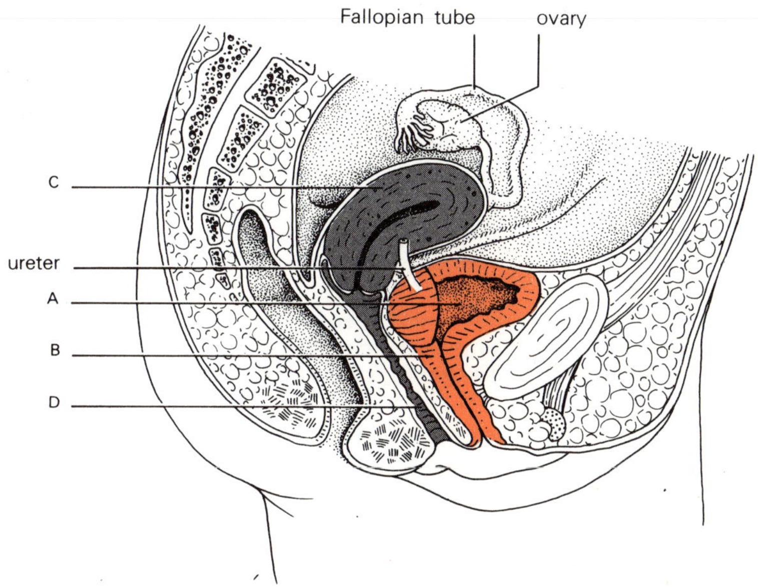

FIGURE 8·2

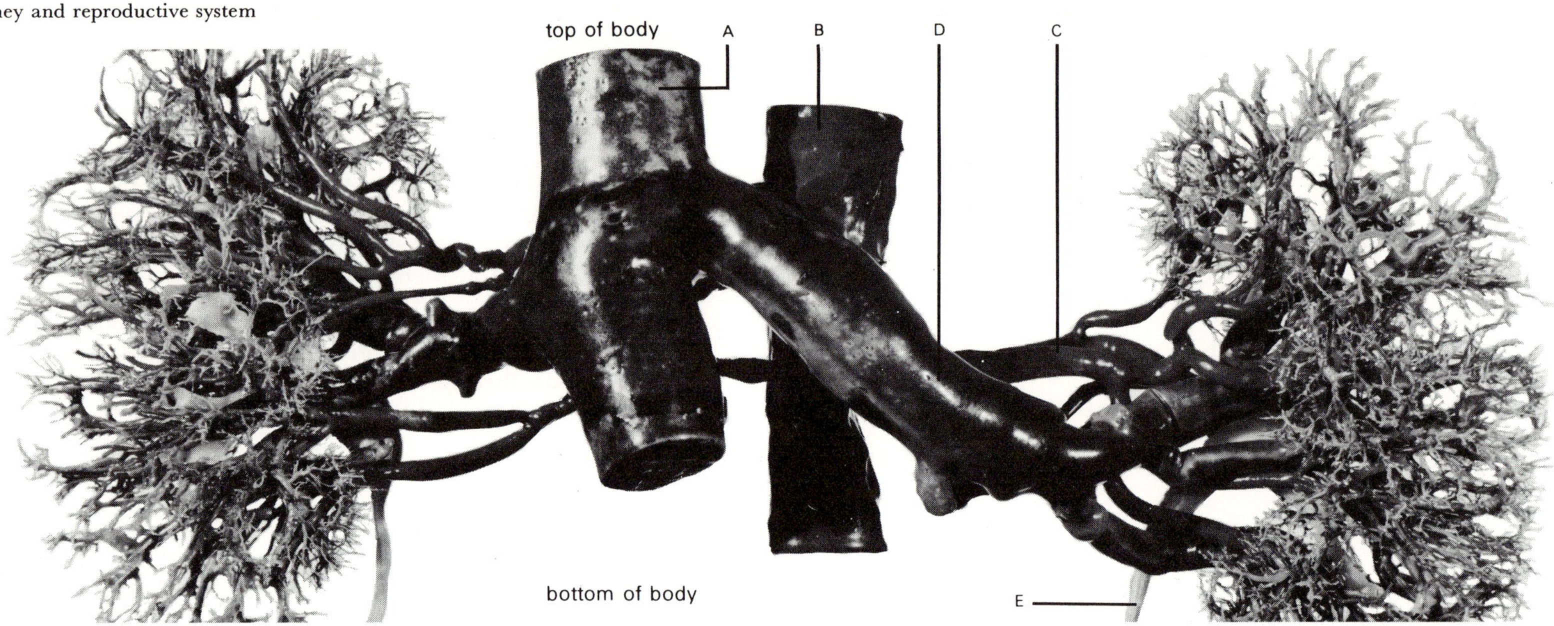

FIGURE 8·3

145

FIGURE 8.3 is a corrosion preparation of the human kidney and the blood vessels serving it. It is seen from the front. It was made by injecting coloured plastic into the blood vessels and ureter of a corpse. The plastic set, after which all other tissues of the kidney and the blood vessels were dissolved away, leaving this cast.

A The structures labelled A, B, C **and** D **are the aorta, the renal vein, the posterior vena cava (the main vein returning blood to the heart from all levels below the heart), and the renal artery.**

Say, giving your reason, which structure you think is which.

B About 4·5l of blood are pumped out of each ventricle per minute in a resting adult. About 1·2l of blood enters the two kidneys per minute.

i What proportion of the output of the ventricles passes through the two kidneys in the resting adult?

ii What does this indicate about the role of the kidneys in regulating the composition of the blood plasma and so of the immediate environment of all the cells of the body?

C From FIGURE 8.3, **what in general can be said about the distribution of the branches of the renal artery and the renal vein?**

D What is the structure labelled E **in** FIGURE 8.3 **?**

What happens to this structure inside the kidney?

146

FIGURE 8.4 is a drawing of a human kidney sliced through the middle. (The adult kidney measures about 130 mm × 60 mm.) It shows more detail of the branching of the blood vessels. Since the kidney is in effect a blood filter, the arrangement of the vessels inside it is important, and is very precise.

There is an outer zone (which is darker) called the *cortex* and an inner (paler) zone called the *medulla*. The cortex is smooth and uninterrupted, but the medulla consists of about 8 to 18 cone-shaped structures, each called a *pyramid*. They empty at the centre of the kidney into branches of the ureter.

In FIGURE 8.4, which is flat, the pyramids look roughly triangular. FIGURE 8.5a is a cut-away stereogram of one pyramid. FIGURE 8.5b

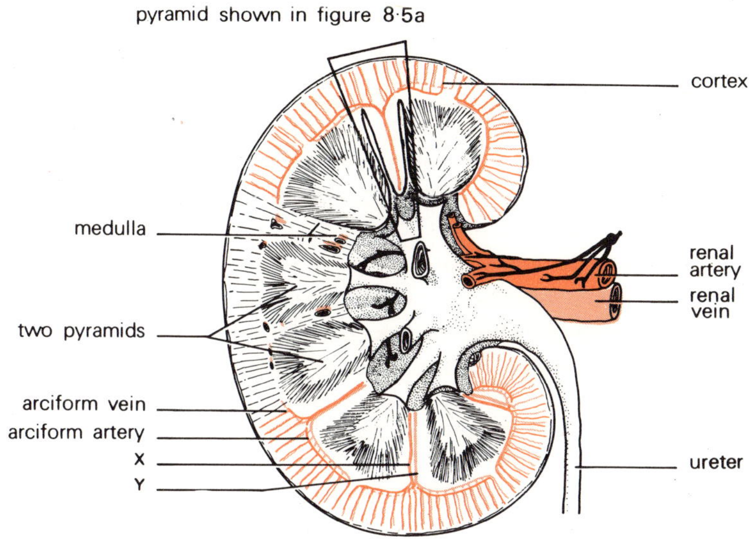

FIGURE 8·4

is a drawing on one of the million kidney filters (nephrons).

A **Study** FIGURE 8.4 **and** 8.5a **carefully and make sure that you can fit** FIGURE 8.5a **into** FIGURE 8.4.

The renal artery divides forming branches which run out between the pyramids towards the surface of the kidney. One of these has been marked X. **Describe as carefully as you can the ways in which this vessel branches until it eventually returns as the vein marked** Y.

B **The arciform arteries mark the boundary between the cortex and the medulla. From** FIGURE 8.5a, **which structures are present in the cortex which are absent from the medulla?**

C **What are the tubes shown cut across at** A **in** FIGURE 8.5a **?**

D **From** FIGURES 8.5a **and** 8.5b, **write a short description of the appearance and position of the uriniferous tubule, that is the tube leading from the renal corpuscle to the collecting duct.**

E **Write a short description of one of the collecting ducts shown in** FIGURES 8.5a **and** 8.5b.

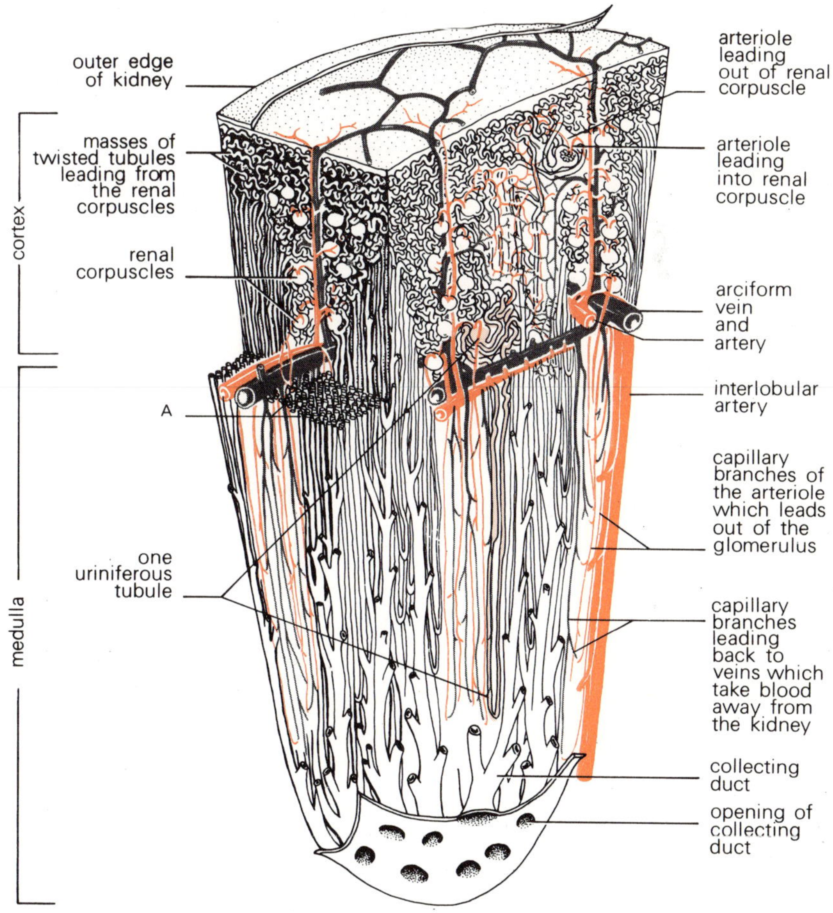

FIGURE 8·5a

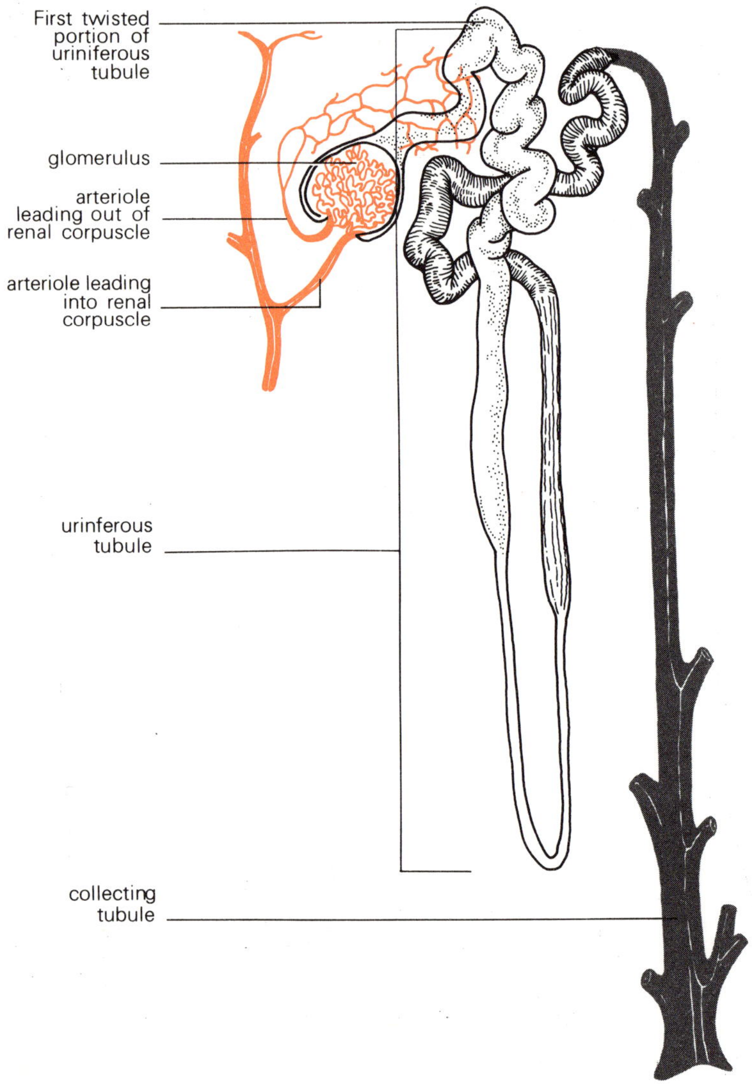

FIGURE 8·5b

We will now look more closely at the renal corpuscle. FIGURE 8.6 is a drawing of one opened up. It is about 0·2 mm in diameter and consists of an outer cup (Bowman's capsule) leading at its base into the uriniferous tubule. Into the cup juts a tuft of about 50 capillary loops served by one arteriole and drained by another. We have already noted where these vessels come from and go to.

A Describe the appearance of

i the cells forming the wall of the cup,

ii the cells forming the first part of the uriniferous tubule,

iii the arterioles leading into and out of the capillary tuft.

B FIGURE 8.7 **is an electron micrograph of a longitudinal section across the part of one of the cells lining the uriniferous tubule seen in** FIGURE 8.6.

What are the structures labelled A **and** B **?**

C The liquid which filters through the capillary tufts is identical with blood plasma except that it contains few, if any, proteins. Blood plasma contains several proteins, including albumin, that make it thicker (more viscous) than water.

When the filtrate leaves the first twisted portion of the uriniferous tubule (see FIGURE 8.5b**), it has been considerably changed. All the glucose and amino acids it contained have disappeared. They have been returned to the blood.**

i What is the purpose of extracting glucose and amino acids from the filtrate?

ii What path must the glucose and amino acids have followed in order to get from the tubule into the blood? To answer this question fully you will have to refer again to FIGURE 8.5a.

D i The proteins in the blood plasma contribute a lot to its osmotic pressure. What, then, will be the osmotic pressure of the newly-filtered liquid in the cup compared with that of plasma?

ii Why, then, is it strange that glucose and amino acids from the filtrate should be completely removed and returned to the blood in the first twisted portion of the uriniferous tubule?

iii What do the structures labelled B **in** FIGURE 8.7 **pro-**

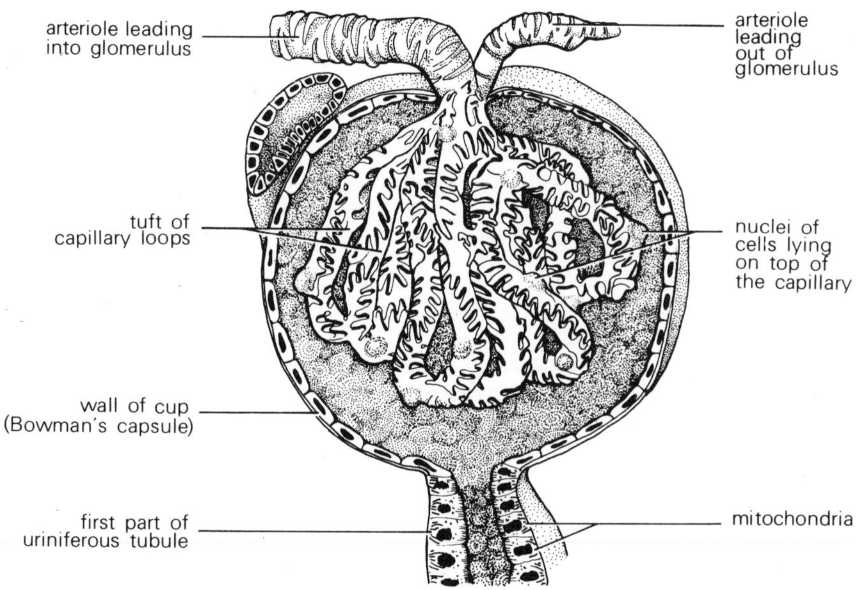

FIGURE 8·6

vide for the cell? What is the significance of the fact that a large number of these structures occur in the cells of the first twisted part of the uriniferous tubule?

iv **What effect will removal of glucose and amino acids from the filtrate have on its osmotic pressure?**

v **Each human kidney contains about one million renal corpuscles. About 60 cm^3 of filtrate are formed in them per minute, passing through the million capillary tufts into the cups.**

The liquid then passes into the first twisted portion of the million uriniferous tubules. Only 10 cm^3 per minute arrive at the far ends of these twisted portions. Not only have all the glucose and amino acids been returned to the blood, but 50 cm^3 per minute of water also.

What mechanism have we discussed in this question that explains the removal of this water?

vi **Make a suggestion about the significance of the structures** A **in** FIGURE 8.7.

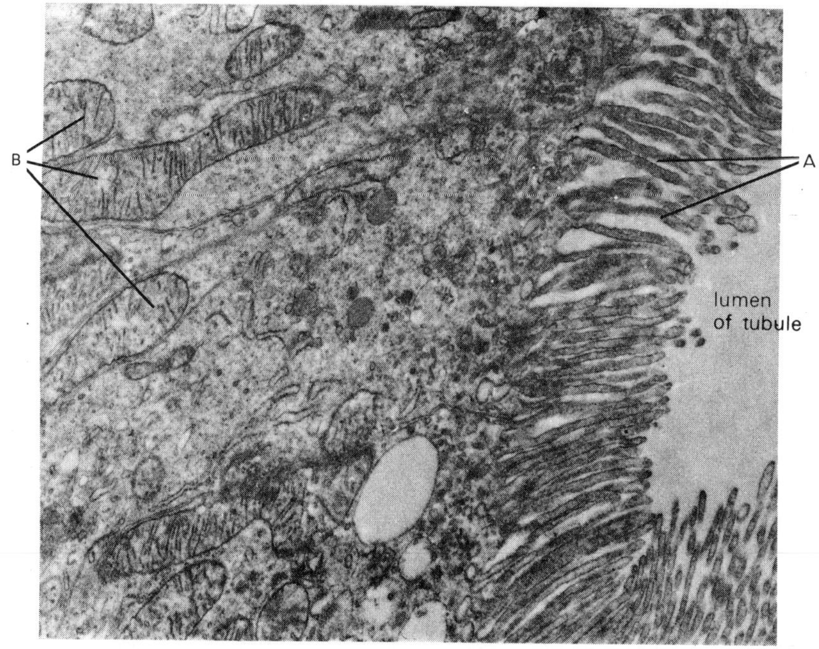

FIGURE 8·7

vii **About 20 cm^3 of treated filtrate arrive at the far ends of the first twisted portions of all the uriniferous tubules in both kidneys per minute. On average a man passes about one litre of urine per day. Urine is produced solely by treatment of the filtered plasma in the uriniferous tubules. What do these facts tell you about the function of the rest of the uriniferous tubule, that is the part beyond the first twisted portion?**

viii **Desert mammals such as the kangaroo rat (see page 55) can produce very concentrated urine, ten times as concentrated as that of an aquatic mammal such as a beaver. When the kidneys of this animal are examined it is found to have a very deep cortex. The straight looped parts of the uriniferous tubules reach to the bottom of it and are thus long. The beaver's kidney, by contrast, has a shallow cortex and the looped parts of the uriniferous tubules are short.**

What light does your answer to the previous question vii throw on the significance of these facts?

148

The liquid which filters through the tufts of capillaries in the renal corpuscle is identical with blood plasma except that it contains few or no proteins.

Plasma filters through the capillaries because it is under strong pressure (about 70 mm Hg) from the heart. The work done in making the filtrate is done by the heart.

In looking at FIGURE 8.6 you must remember that the space between the tuft of capillaries and the wall of the cup is full of filtrate, as is the uriniferous tubule.

A **What effect will this have on the rate of filtration?**

B **Remember that the plasma in the capillary tufts contains blood proteins and that the filtrate bathing the tufts does not. What effect will this fact have on the rate of filtration?**

C **What, then, is the minimum level of blood pressure essential to filtration taking place?**

D **The kidneys lie only a few centimetres from the heart. The renal arteries are short (see** FIGURE 8.3 **p 132). How are these facts related to your answer to C ?**

E **How does each of the following experimental observations fit in with the picture you have just obtained of the forces operating in the formation of filtrate from the capillary tuft?**

I **Lowering the blood pressure in the whole body by stimulating the vagus nerves (which reduces the rate and strength of heartbeat) reduces the flow of urine.**

II **Constriction of the arteriole leading into the renal corpuscle causes reduction in the amount of filtrate formed.**

III **Constriction of the arteriole leading out of the renal corpuscle causes an increase in the amount of filtrate formed.**

IV **Partial blocking of the ureter by constricting it but not blocking it altogether causes a pile-up of liquid all the way back to the renal corpuscle. This reduces the flow of urine.**

149

We have discovered that the filtered plasma is treated by the first twisted portion of the uriniferous tubule so that all the glucose, all the amino acids and about 83% of the water it contains are removed from it. (From the 120 cm³ of filtrate formed per minute, 100 cm³ of water are reabsorbed in the first twisted part of the tubules.)

While this fraction of water is *always* removed here, a variable amount is removed (taken back into the blood stream) in the remaining part of the tubules. If the body is short of water, little or none gets past this part; if the body has an abundance of water (e.g. after drinking several pints of beer), a lot gets past this part.

A FIGURE 8.8 **shows in line** A **the effect on a man's urine production of drinking one litre of water. Urine was collected at half-hourly intervals. Before drinking the litre of water, his output was about 59 cm³ per half hour.**

i **Describe his urine output over the four-hour period after drinking.**

ii **How long did it take for him to eliminate as urine all the water he had drunk?**

B **On the following day the man repeated the experiment but this time drank one litre of 0·9% sodium chloride solution (line** B **of** FIGURE 8.8). **This is** isotonic **with blood plasma, i.e. has the same osmotic strength as blood plasma.**

i **What effect did drinking this liquid have on urine production over the following four hours?**

ii **What happened to the litre of saline over this period?**

iii **What do the results of this experiment indicate about the effect of the kidneys on the volume of the blood plasma?**

iv **What does a comparison of the results of the two experiments indicate about the effect of the kidneys on the osmotic pressure of the blood plasma?**

150

Elimination of one litre of water in the previous experiment occurred because the portions of the uriniferous tubules past the first twisted portion did *not* absorb water from the dilute filtrate passing over them. When the extra litre of water had passed out of the kidneys, these parts of the tubules reverted to absorbing water again.

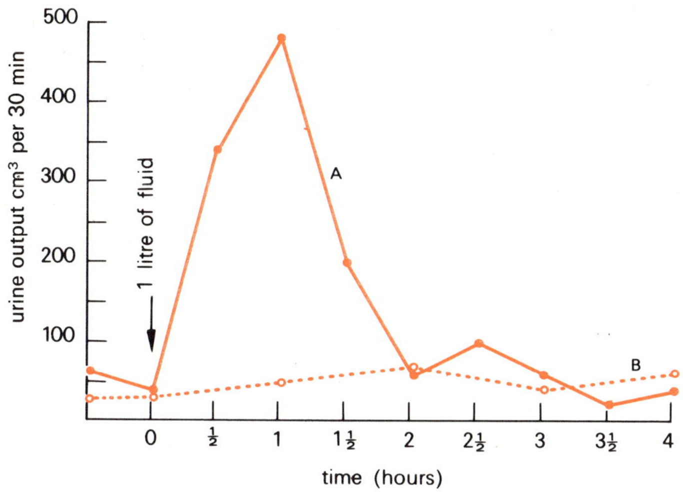

FIGURE 8·8

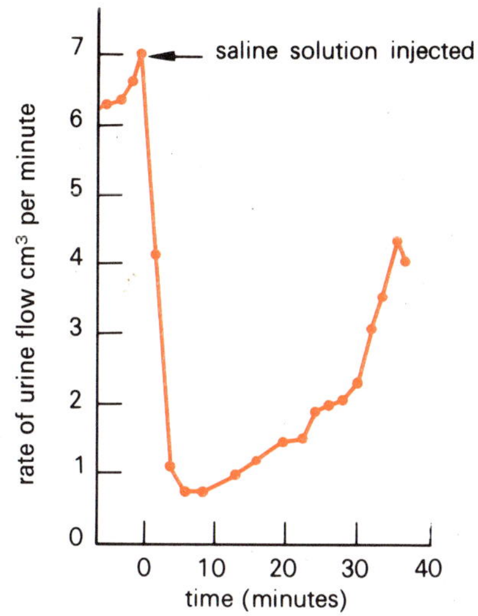

FIGURE 8·9

There must, therefore, be some control mechanism which 'tells' these parts of the tubules when to absorb water from the filtrate passing over them, and when not to.

One possibility is that the cells of the tubule are themselves sensitive to the osmotic pressure of the liquid passing over them. In other words, the osmotic pressure of the filtrate 'tells' the cells directly whether to absorb water from the filtrate or not.

However, the following experiment indicates that this is not the case and gives a good clue as to the real controlling mechanism.

10 cm^3 of a solution of sodium chloride were injected into the carotid artery of a dog. This artery passes up the neck and serves the brain. If you look back to FIGURE 3.7 (p 46), you will remember how some of its branches touch the thalamus and hypothalamus.

The strength of the solution was 2·5%. A 0·9% solution of sodium chloride has the same osmotic strength as blood plasma. It had previously been found that if this much 2·5% sodium chloride solution was injected into a *vein*, it had no effect at all on the amount of urine produced by the dog. From the vein it passed back to the heart and so became diluted by the rest of the animal's blood.

But when injected into the carotid artery, the solution passed directly on in the blood stream and reached the brain before there had been time for much dilution.

The effect of this injection on the rate of urine production is shown in FIGURE 8.9.

A **What effect did the injection have on the rate of urine production during the first three minutes after injection?**

B **Where is it that the cut-back in urine production occurred?**

C **It is a fair guess that injection of sodium chloride solution had its effect because of its high osmotic pressure. Similar injections of solutions of other chemicals (glucose, sucrose, sodium sulphate) of equal osmotic strengths all had the same effect on urine output.**

i **How would this suggestion be in keeping with the results shown in FIGURE 8.8, line A?**

ii **When the salt solution was injected into the carotid artery it had a very different effect on urine production compared with when it was injected into a vein. Explain how this difference shows that the osmotic effect does not act directly on the cells of the uriniferous tubules, causing them to change the amount of water they absorb from the filtrate passing over them.**

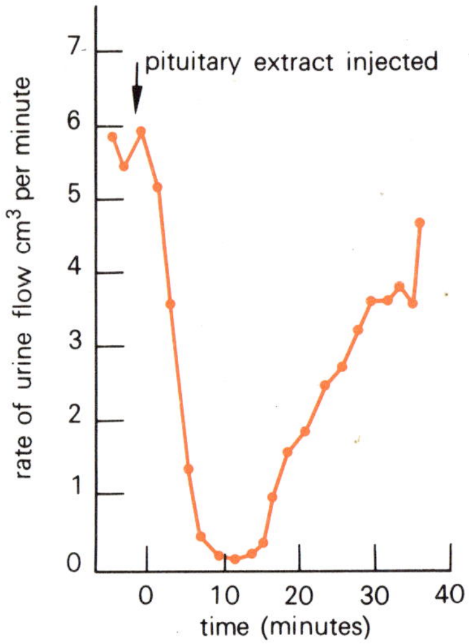

FIGURE 8·10

151

Another experiment was performed with the same dog. An extract from the pituitary gland of another dog was prepared and was injected into a vein of the experimental dog. The effect on its rate of urine production is shown in FIGURE 8.10.

A **How does the pattern of change in urine production in this experiment compare with that in the previous one, when 2·5% sodium chloride solution was injected into the dog's carotid artery?**

B **Something in the pituitary extract must have been affecting the kidney in order to produce the result shown in** FIGURE 8.10.

i **What sort of substance is this likely to be? Give the reason for your answer.**

ii **What effect must it have had on the kidney to produce this result?**

C **What effect might injection of the salt solution into the carotid artery have on the pituitary body that would cause**

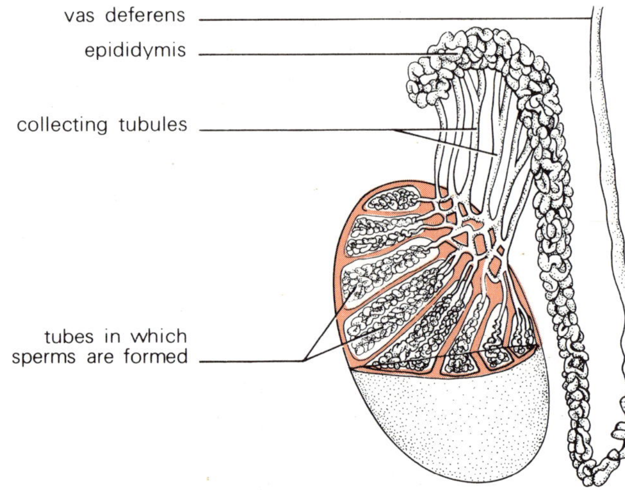

FIGURE 8·11

the fall in urine output shown in FIGURE 8.9 **?**

152

Inside the testis is a mass of twisted tubes; sperms are produced by the cells lining the tubes.

The tubes lie in compartments (FIGURE 8.11), of which there are about 300. Each contains three twisted tubes; and each tube is about 0·8 m long. They lead into the *epididymis*, a twisted tube some six metres long and 1 mm wide, where the sperms are stored. The epididymis has muscular walls and can move the sperms along by peristaltic contractions.

A **From** FIGURE 8.1 **(p 131), describe the path travelled by sperms from their point of origin to the tip of the penis, where they are ejaculated.**

B **Look back to** FIGURE 8.2 **(p 131). What is the difference in the positions of the testes and the ovaries?**

C **The testes enter the scrotum of the developing baby just before or at birth. Sometimes they do not descend, and**

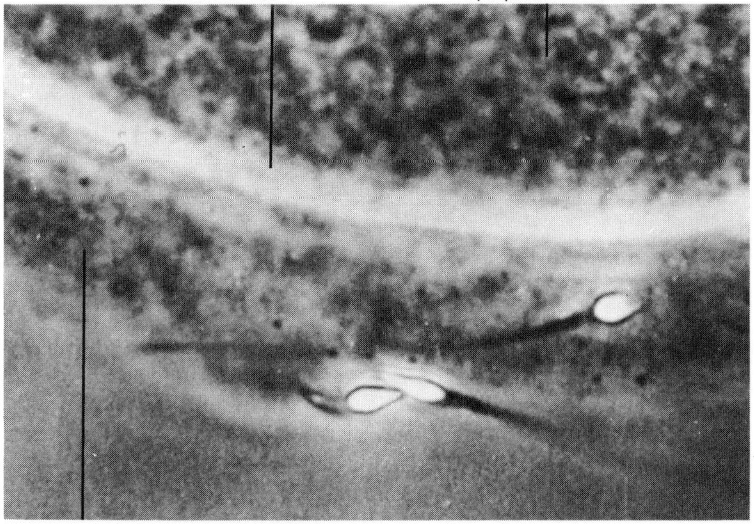

FIGURE 8·12

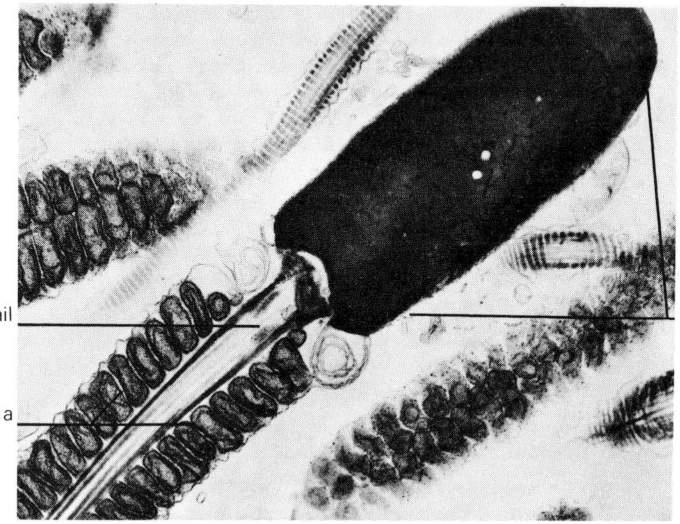

FIGURE 8·13

in that case the testis does not produce sperms. It has been shown in cats that if the scrotum is insulated in such a way as to keep it at the animal's internal body temperature, sperms do not develop. The skin of the human scrotum contains many muscle fibres. If the body is warm, these relax and the testes hang further away from the groin. If the body is cold, they contract, drawing the testes close to the groin.

What light do these facts throw on the difference in position between the testes and the ovaries in man?

D **How is the arrangement of the urethra different in the woman from the man?**

E **A human ejaculation is about 4 cm³ of semen, 10% of which is occupied by the sperms themselves. There are about 300 million sperms in one ejaculation.**

A woman produces one egg per month from the age of about 14 to about 45. How, then, does the number of male gametes compare with the number of female gametes produced?

F FIGURE 8.12 **shows part of a single human ovum, which is about 0·1 mm in diameter. It is surrounded by a thin cluster of other cells which feed it. Three sperms are shown, one about to penetrate the egg.**

FIGURE 8.13 **is a longitudinal section of a bat sperm (an electron micrograph).**

All mammalian sperms and ova have a similar structure.

i **List as many differences as you can between the appearance of the sperm and the appearance of the ovum.**

ii **Explain the presence of the mitochondria in the tail of the sperm.**

iii **Try to explain the difference between the numbers of sperms and ova produced.**

iv **Fish and amphibia release their gametes into the water in which they live. Usually the male and female animals are close to each other and release the gametes at the same time, thereby increasing the chance of a sperm finding an ovum.**

In man, the erect penis is inserted into the vagina and semen is ejaculated to the top of the vagina. Birds and reptiles also have methods of copulating, sometimes involving a penis.

How is copulation a vital adaptation to life on the land as compared with life in the water?

G **The act of ejaculation can be divided into two parts. First, a second or two before ejaculation, the sperms, which are immobile in the epididymis, are moved up the vas deferens by peristaltic contractions. They collect in the part of the urethra which passes through the prostate gland (FIGURE 8.1 p 131). This becomes distended. Secondly, strong muscles surrounding the base of the penis contract several times, squeezing the urethra and all the glands associated with it (that is, the seminal vesicles, the prostate gland and Cowper's glands). This causes the secretions of the glands to be mixed with the sperms, the whole mixture (semen) being shot out of the top of the penis.**

The liquid added by the seminal vesicles contributes about 60% of the volume of the semen, and it contains a high concentration of the sugar fructose.

i **What do you suppose is the function of the fructose in the semen?**

ii **The sperms have scanty cytoplasm. How might this be related to the fact that they do not move until they have been mixed with the secretion of the glands and are part of the semen?**

153

FIGURE 8.14 shows a side view of a five-week old developing baby in the womb. Look back to FIGURE 8.2 (p 131) and identify the womb. You can see that the baby is attached to the wall of the womb by the *umbilical cord* and the *placenta* (FIGURE 8.15). The cord contains two *umbilical arteries*, which take blood from the baby to the placenta, and an *umbilical vein* which takes blood back from the placenta to the baby. FIGURE 8.16 is a transverse section of the cord showing these three vessels (the arteries have contracted).

The baby receives all its foods, its oxygen and other chemicals such as antibodies, from its mother's blood.
FIGURE 8.17 shows a small part of the placenta in detail.

A **i** **Describe the arrangement of the blood vessels and spaces on the baby's and on the mother's side of the placenta.**

ii **Describe the course of the circulation through the placenta on the baby's and on the mother's side.**

B **The mature placenta is about the size of a dinner plate and has a volume of about 500 cm³. Of this, about 350 cm³ is occupied by the capillary tufts on the baby's side (the villi). Therefore the spaces surrounding the villi have a total volume of about 150 cm³.**

The mother's blood circulates through the placenta at a rate of about 500 cm³ per minute.

i **How frequently is the maternal blood in the placenta replaced?**

ii **I** **The total length of all the capillaries in the villi is something like 50 km. Their surface area is about 14 m².**

II **The baby's blood flows through the villi at a rate of about 400 cm³ per minute, that is almost all of its blood flows through the placenta each minute.**

What do these two facts indicate about the activity of the placenta?

C **Glucose and oxygen are both more concentrated in the mother's blood than in the developing baby's blood. What will happen to these substances, therefore, in the placenta?**

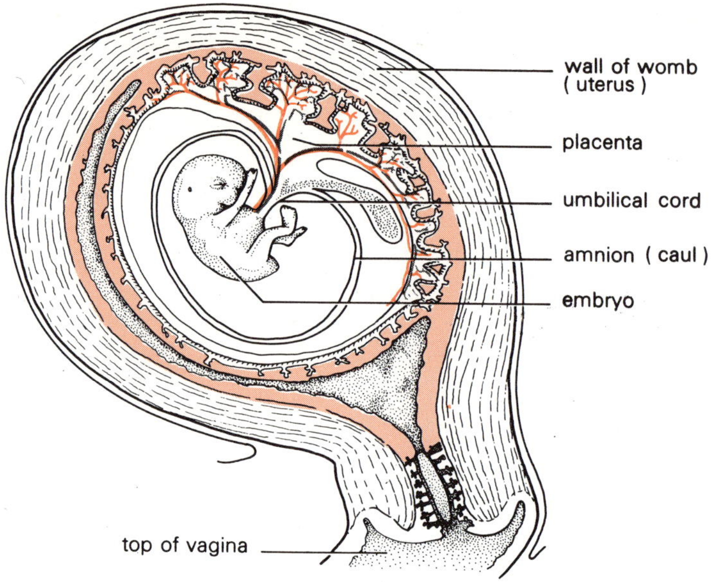

FIGURE 8·14

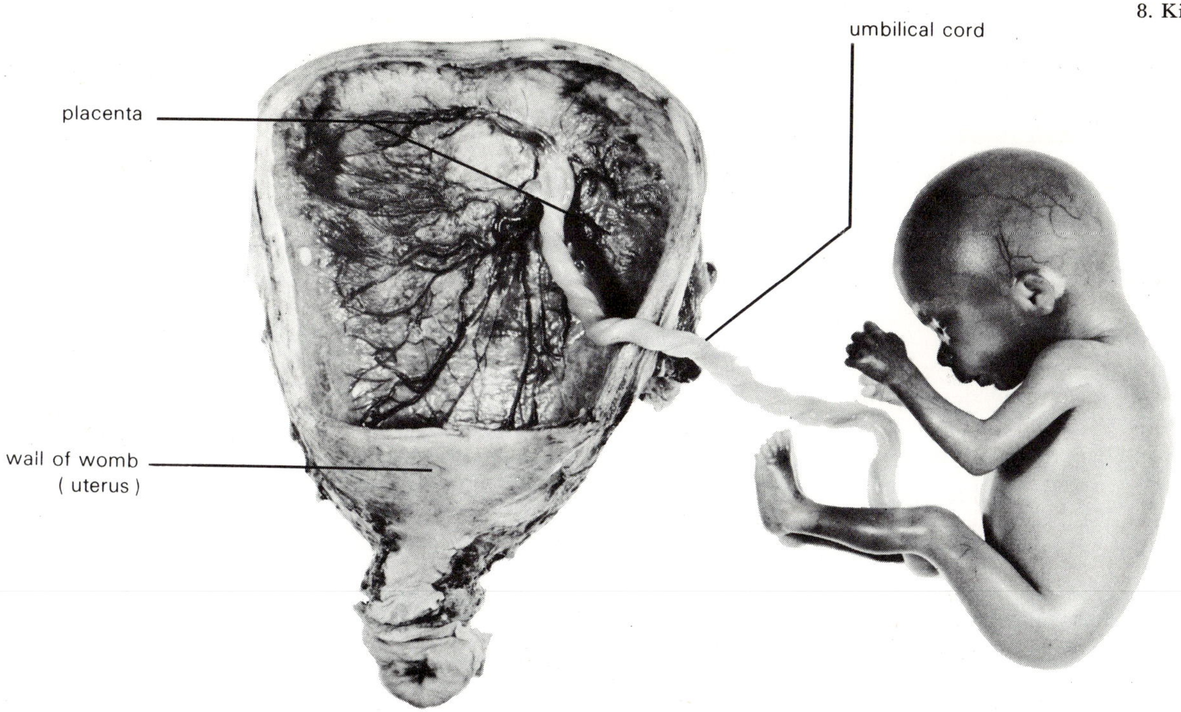

FIGURE 8·15

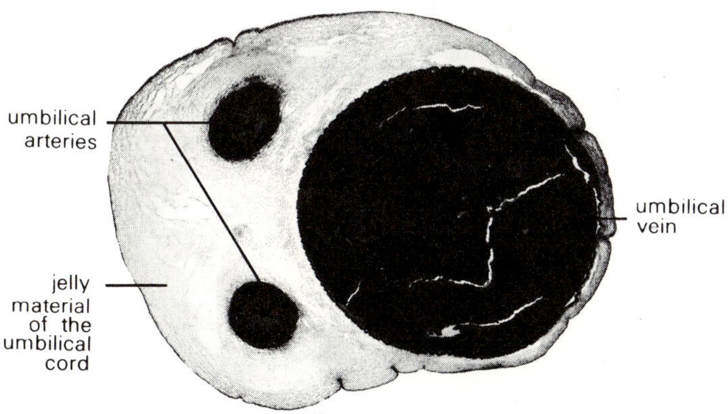

FIGURE 8·16

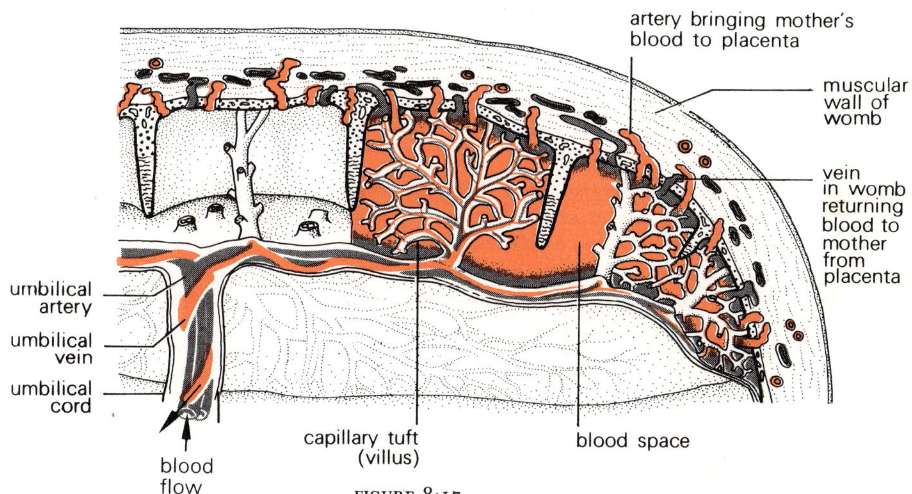

FIGURE 8·17

D **The concentration of amino acids in the developing baby's circulation is always slightly higher than that in the mother's circulation. Yet injection of amino acids into the mother's circulation increases their concentration in the baby's blood. Why is this fact surprising?**

E **What are the main waste products produced by the developing baby that move in the other direction, from the baby's blood in the villi to the mother's blood in the spaces surrounding the villi?**

154

The next two questions are concerned with the way in which blood flows round the body of the developing baby. This circulation is considerably different from that of the adult, and immediately the baby is born and starts to breathe, its circulation changes and adjusts towards the adult pattern.

First we will recall the adult circulation (FIGURE 8.18).

A **i** **Which vessels empty into the right auricle? From where does the blood in them come, and what is its state of oxygenation?**

ii **Which vessels empty into the left auricle? From where does the blood in them come, and what is its state of oxygenation?**

iii **Which vessel leads out of the right ventricle? Where does it lead, and what is the state of oxygenation of the blood in it?**

iv **Which vessel leads out of the left ventricle? Where does it lead, and what is the state of oxygenation of the blood in it?**

B **The circulation in a baby in the womb is shown in** FIGURE 8.19. **The main differences between it and the circulation in the adult are caused by two facts:**

I **the baby does not breathe in the womb. Its alveoli are crumpled and do not expand till the first breath. It makes no breathing movements.**

II **the baby's blood circulates not only round its own body but also to the placenta in the umbilical arteries and back again in the umbilical vein.**

Apart from the umbilical vessels, there are two other short vessels in the developing baby which are absent in the adult, the ductus venosus **and the** ductus arteriosus.

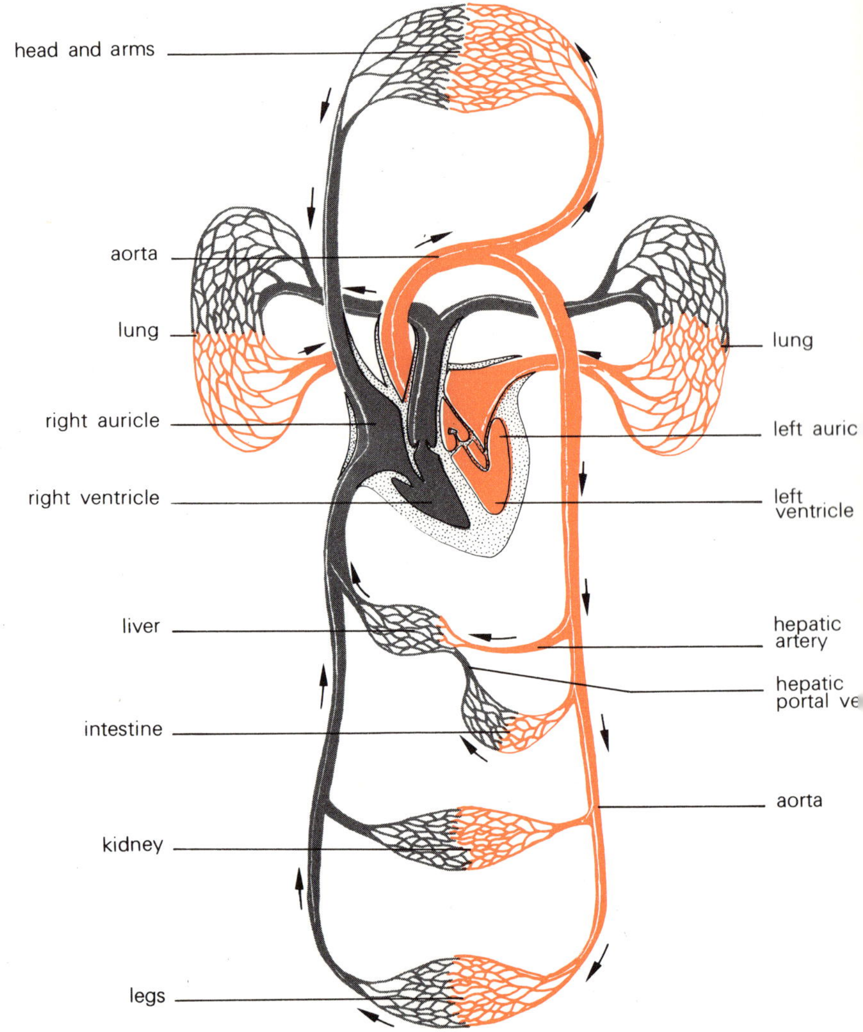

FIGURE 8·18

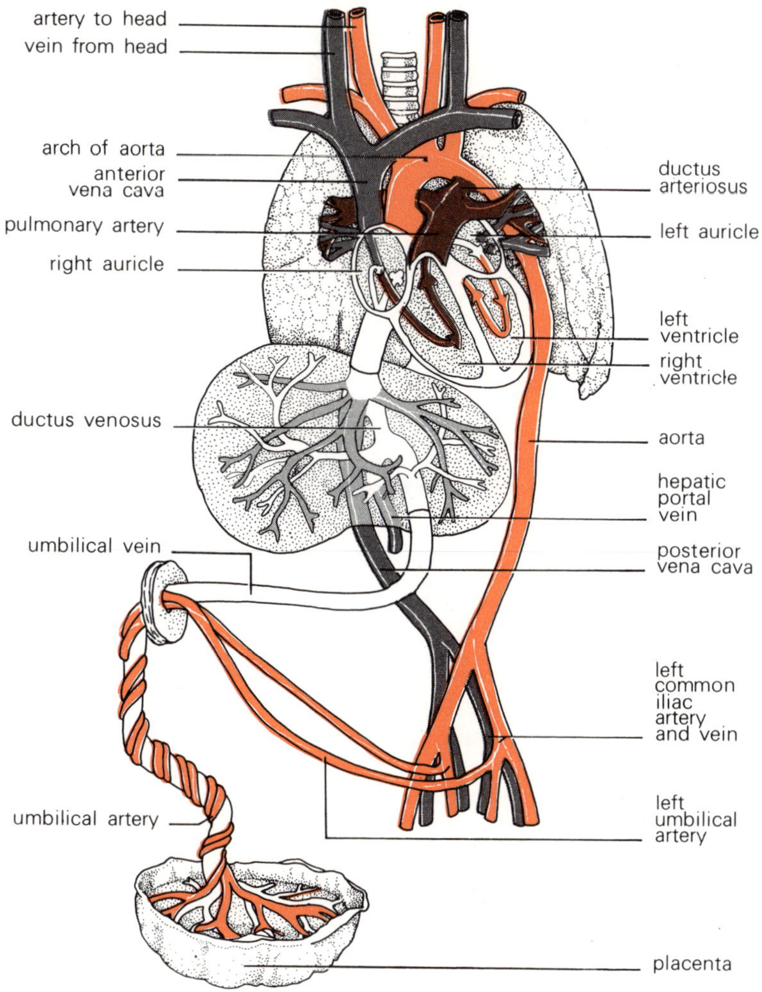

FIGURE 8·19

Find them in FIGURE 8.19. **Each of them is wide and short and connects two other vessels.**

- **i** **Which vessels are connected by the ductus venosus?**
- **ii** **Which vessels are connected by the ductus arteriosus?**
- **iii** **From** FIGURE 8.19, **how does the blood returning from the placenta in the umbilical vein get back to the heart?**
- **iv** **How will the presence of the ductus arteriosus in the unborn baby make the distribution of blood leaving its right ventricle different from the distribution in the adult?**
- **v** **Some blood passes to the baby's lungs along its pulmonary arteries and back to its left auricle along its pulmonary veins. The lungs are not functioning. Why do they need a blood supply?**

C **Another difference between the baby in the womb and the adult is that whereas in the adult the auricles are separated from each other by a wall in the middle of the heart** (FIGURE 8.18), **in the unborn baby there is a large flap in this wall which allows blood to pass from the right auricle into the left auricle.**

How will the distribution of the blood leaving the unborn baby's right auricle differ from the adult circulation?

D **To summarize, the vessels leading into and out of the heart of the unborn baby are the same as those in the adult. The two differences in the heart are:**

- **I** **the right auricle opens by a flap into the left auricle;**
- **II** **some of the blood leaving the right ventricle passes into the aorta via the ductus arteriosus.**

Taking the chambers of the unborn baby's heart in turn (right auricle, left auricle, right ventricle, left ventricle), state:

- **i** **for the auricles, where the blood entering the chamber has come from and what its state of oxygenation is;**
- **ii** **for the ventricles, where the blood leaving is going to, and what its state of oxygenation is;**
- **iii** **how these two facts differ from the arrangement in the adult circulation.**

155

We will now see how the circulation changes at birth and assumes the adult course.

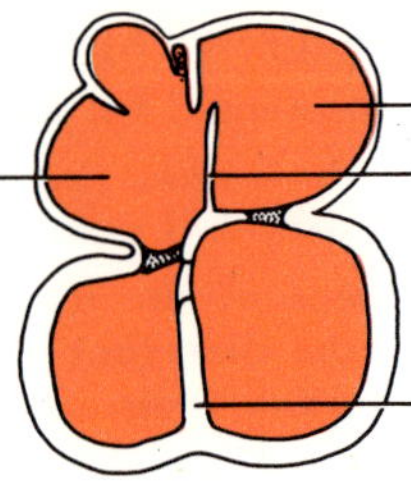

FIGURE 8·20

At birth, the umbilical cord is tied off in two places and cut between them. Apart from this, the umbilical arteries and veins themselves contract, preventing any flow of blood away from or towards the baby.

A **What effect will this have on the blood entering the right auricle?**

B **At the same time, the ductus arteriosus narrows considerably, What effect will this have on the distribution of blood leaving the right ventricle?**

C **At the same time the baby takes its first breath. The lungs expand. The folded-up alveoli open out and fill with air. The arteries, veins, arterioles, venules and capillaries of the lung all enlarge suddenly.**

What effect will this have on the distribution of blood leaving the right ventricle?

D FIGURE 8.20 **illustrates the flap which, until birth, has allowed blood to flow from the right auricle into the left auricle. You can see that the large lower part of the flap overlaps a shorter part which projects downwards.**

This flap must now close so that the blood flowing through the right side of the heart is kept separate from that flowing through the left side.

This change happens because the pressure of blood in the left auricle rises suddenly at birth, pushing the flap and closing it.

How does the expansion of the lungs and the constriction

AGE (in lunar months)	WEIGHT (g)	INCREASE IN WEIGHT PER MONTH
3	20	
4	120	
5	300	
6	635	
7	1220	
8	1700	
9	2240	
10	3250 (i.e., approx. 7 pounds)	

Note that all these measurements show marked variation depending on genetic, nutritional and general environmental factors (e.g., age of mother and number of preceding pregnancies).

FIGURE 8·21

of the ductus arteriosus cause a sudden rise in pressure in the left auricle, thus closing the flap?

156

FIGURE 8.12 shows the body weight of an average developing human baby at the end of the 3rd to the 10th lunar months (a lunar month is four weeks, or 28 days).

A **Copy the table and fill in the right-hand column, showing the increase in weight per month.**

B **What proportion of the baby's weight was acquired in (i) its last two lunar months of its development (ii) the first five lunar months of its development?**

C **All the organs are present in the baby at the end of the eighth lunar months. From** FIGURE 8.21, **what do you suppose is happening to the baby in its last two lunar months in the womb?**

D I **In the last two months of development a great deal of fat is deposited under the baby's skin.**

II **A baby born two months before its time (a premature baby) can usually be kept alive, but it has to live in an incubator for several weeks.**

How are these two facts related?